# Calculus

*by*
*Bernard V. Zandy, B.S., M.A.*

Series Editor
Jerry Bobrow, Ph.D.

LINCOLN, NEBRASKA 68501

FIRST EDITION



ISBN 0-8220-5312-8

# CONTENTS

## CONTENTS

# REVIEW TOPICS

Certain topics in algebra, geometry, analytical geometry, and trigonometry are essential in preparing to study calculus. Some of them are briefly reviewed below.

## Interval Notation

The set of real numbers ($R$) is the one that you will be most generally concerned with as you study calculus. This set is defined as the union of the set of rational numbers with the set of irrational numbers. Interval notation provides a convenient abbreviated notation for expressing intervals of real numbers without using inequality symbols or set-builder notation.

The following lists some common intervals of real numbers and their equivalent expressions, using set-builder notation:

$$(a,b) = \{x \in R: a < x < b\}$$

$$[a,b] = \{x \in R: a \leq x \leq b\}$$

$$[a,b) = \{x \in R: a \leq x < b\}$$

$$(a,b] = \{x \in R: a < x \leq b\}$$

$$(a,+\infty) = \{x \in R: x > a\}$$

$$[a,+\infty) = \{x \in R: x \geq a\}$$

$$(-\infty,b) = \{x \in R: x < b\}$$

$$(-\infty,b] = \{x \in R: x \leq b\}$$

$$(-\infty,+\infty) = \{x \in R\}$$

Note that an infinite end point ($\pm\infty$) is never expressed with a bracket in interval notation because neither $+\infty$ nor $-\infty$ represents a real number value.

## Absolute Value

The concept of absolute value has many applications in the study of calculus. The absolute value of a number $x$, written $|x|$ may be defined in a variety of ways. On a real number line, the absolute value of a number is the distance, disregarding direction, that the number is from zero. This definition establishes the fact that the absolute value of a number must always be nonnegative—that is, $|x| \geq 0$.

A common algebraic definition of absolute value is often stated in three parts, as follows:

$$|x| = \begin{cases} x, & x > 0 \\ 0, & x = 0 \\ -x, & x < 0 \end{cases}$$

Another definition that is sometimes applied to calculus problems is

$$|x| = \sqrt{x^2}$$

or the principal square root of $x^2$. Each of these definitions also implies that the absolute value of a number must be nonnegative.

## Functions

A function is defined as a set of ordered pairs $(x,y)$, such that for each first element $x$, there corresponds one and only one second element $y$. The set of first elements is called the domain of the function, while the set of second elements is called the range of the function. The domain variable is often referred to as the independent variable, and the range variable is referred to as the dependent variable. The notation $f(x)$ is often used in place of $y$ to indicate the value of the function $f$ for a specific replacement for $x$ and is read "$f$ of $x$" or "$f$ at $x$."

Geometrically, the graph of a set of ordered pairs $(x,y)$ represents a function if any vertical line intersects the graph in, at most, one point. If a vertical line were to intersect the graph at two or more points, the set would have one $x$ value corresponding to two or more $y$ values, which clearly contradicts the definition of a function. Many of the key concepts and theorems of calculus are directly related to functions.

**Example 1:** The following are some examples of equations that are functions.

(a) $y = f(x) = 3x + 1$

(b) $y = f(x) = x^2$

(c) $y = f(x) = |x| - 5$

(d) $y = f(x) = -3$

(e) $y = f(x) = \dfrac{x - 3}{x^2 + 4}$

(f) $y = f(x) = \sqrt[3]{2x + 9}$

(g) $y = f(x) = \dfrac{6}{x}$

(h) $y = \tan x$

(i) $y = \cos 2x$

**Example 2:** The following are some equations that are not functions; each has an example to illustrate why it is not a function.

(a) $x = y^2$. If $x = 4$, then $y = 2$ or $y = -2$.

(b) $x = |y + 3|$. If $x = 2$, then $y = -5$ or $y = -1$.

(c) $x = -5$. If $x = -5$, then $y$ can be any real number.

(d) $x^2 + y^2 = 25$. If $x = 0$, then $y = 5$ or $y = -5$.

(e) $y = \pm\sqrt{x + 4}$. If $x = 5$, then $y = +3$ or $y = -3$.

(f) $x^2 - y^2 = 9$. If $x = -5$, then $y = 4$ or $y = -4$.

## Linear Equations

A linear equation is any equation that can be expressed in the form $ax + by = c$, where $a$ and $b$ are not both zero. Although a linear equation may not be expressed in this form initially, it can be manipulated algebraically to this form. The slope of a line indicates whether the line slants up or down to the right or is horizontal or vertical. The slope is usually denoted by the letter $m$ and is defined in a number of ways:

$$\begin{aligned} m &= \frac{\text{rise}}{\text{run}} \\ &= \frac{\text{vertical change}}{\text{horizontal change}} \\ &= \frac{y \text{ value change}}{x \text{ value change}} \\ &= \frac{\Delta y}{\Delta x} \\ &= \frac{y_1 - y_2}{x_1 - x_2} \\ &= \frac{y_2 - y_1}{x_2 - x_1} \end{aligned}$$

Note that for a vertical line, the $x$ value would remain constant, and the horizontal change would be zero; hence, a vertical line is said to have no slope or its slope is said to be nonexistent or undefined. All nonvertical lines have a numerical slope with a positive slope indicating a line slanting up to the right, a negative slope indicating a line slanting down to the right, and a slope of zero indicating a horizontal line.

**Example 3:** Find the slope of the line passing through $(-5,4)$ and $(-1,-3)$.

$$m = \frac{y_2 - y_1}{x_2 - x_1}$$
$$= \frac{(-3) - (4)}{(-1) - (-5)}$$
$$= -\frac{7}{4}$$

The line, then, has a slope of $-7/4$.

Some forms of expressing linear equations are given special names that identify how the equations are written. Note that even though these forms appear to be different from one another, they can be algebraically manipulated to show they are equivalent.

Any nonvertical lines are parallel if they have the same slopes, and conversely, lines with equal slopes are parallel. If the slopes of two lines $L_1$ and $L_2$ are $m_1$ and $m_2$, respectively, then $L_1$ is parallel to $L_2$ if and only if $m_1 = m_2$. Two nonvertical, nonhorizontal lines are perpendicular if the product of their slopes is $-1$, and conversely, if the product of their slopes is $-1$, the lines are perpendicular. If the slopes of two lines $L_1$ and $L_2$ are $m_1$ and $m_2$, respectively, then $L_1$ is perpendicular to $L_2$ if and only if $m_1 \cdot m_2 = -1$. You should note that any two vertical lines are parallel and a vertical line and a horizontal line are always perpendicular.

The **general** or **standard form** of a linear equation is $ax + by = c$, where $a$ and $b$ are not both zero. If $b = 0$, the equation takes the form $x =$ constant and represents a vertical line. If $a = 0$, the equation takes the form $y =$ constant and represents a horizontal line.

**Example 4:** The following are some examples of linear equations expressed in general form:

(a) $2x + 5y = 10$
(b) $x - 4y = 0$
(c) $x = -3$
(d) $y = 6$

The **point-slope form** of a linear equation is $y - y_1 = m(x - x_1)$ when the line passes through the point $(x_1, y_1)$ and has slope $m$.

**Example 5:** Find an equation of the line through the point (3,4) with slope −2/3.

$$y - y_1 = m(x - x_1)$$

$$y - 4 = -\frac{2}{3}(x - 3)$$

$$y - 4 = -\frac{2}{3}x + 2$$

$$y = -\frac{2}{3}x + 6$$

$$3y = -2x + 18$$

$$2x + 3y = 18 \quad \text{(general form)}$$

The **slope-intercept form** of a linear equation is $y = mx + b$ when the line has $y$-intercept $(0,b)$ and slope $m$.

**Example 6:** Find an equation of the line that has slope 4/3 and crosses the $y$-axis at $-5$.

$$y = mx + b$$

$$y = \frac{4}{3}x + (-5)$$

$$3y = 4x - 15$$

$$4x - 3y = 15 \quad \text{(general form)}$$

The **intercept form** of a linear equation is $x/a + y/b = 1$ when the line has $x$-intercept $(a,0)$ and $y$-intercept $(0,b)$.

**Example 7:** Find an equation of the line that crosses the $x$-axis at $-2$ and the $y$-axis at 3.

$$\frac{x}{a} + \frac{y}{b} = 1$$

$$\frac{x}{-2} + \frac{y}{3} = 1$$

$$-3x + 2y = 6 \quad \text{(general form)}$$

## Trigonometric Functions

In trigonometry, angle measure is expressed in one of two units: degrees or radians. The relationship between these measures may be expressed as follows: $180^\circ = \pi$ radians. To change degrees to radians, the equivalent relationship $1^\circ = \pi/180$ radians is used and the given number of degrees is multiplied by $\pi/180$ to convert to radian measure. Similarly, the equation 1 radian = $180°/\pi$ is used to change radians to degrees by multiplying the given radian measure by $180/\pi$ to obtain the degree measure.

The six basic trigonometric functions may be defined using a circle with equation $x^2 + y^2 = r^2$ and an angle $\theta$ in standard position with its vertex at the center of the circle and its initial side along the positive portion of the $x$-axis (Figure 1).

The trigonometric functions sine, cosine, tangent, cotangent, secant, and cosecant are defined as follows:

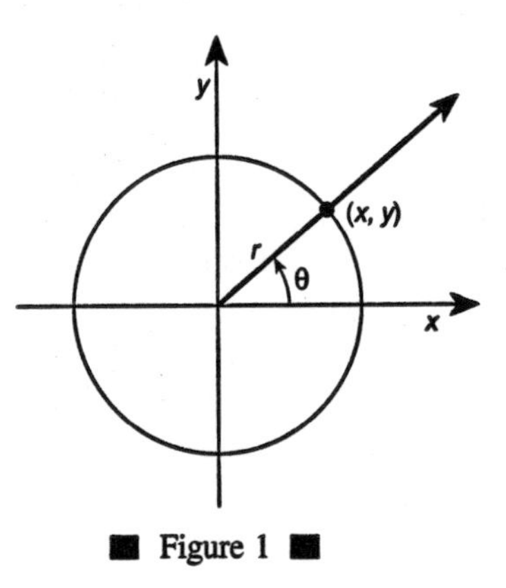


■ Figure 1 ■

$$\sin \theta = \frac{y}{r} = \frac{y}{\sqrt{x^2 + y^2}}$$

$$\cos \theta = \frac{x}{r} = \frac{x}{\sqrt{x^2 + y^2}}$$

$$\tan \theta = \frac{\sin \theta}{\cos \theta} = \frac{y}{x}$$

$$\cot \theta = \frac{\cos \theta}{\sin \theta} = \frac{x}{y}$$

$$\sec\theta = \frac{1}{\cos\theta} = \frac{r}{x} = \frac{\sqrt{x^2+y^2}}{x}$$

$$\csc\theta = \frac{1}{\sin\theta} = \frac{r}{y} = \frac{\sqrt{x^2+y^2}}{y}$$

It is essential that you be familiar with the values of these functions at multiples of 30°, 45°, 60°, 90°, and 180° (or in radians, $\pi/6$, $\pi/4$, $\pi/3$, $\pi/2$, and $\pi$) (Table 1, page 11). You should also be familiar with the graphs of the six trigonometric functions. Some of the more common trigonometric identities that are used in the study of calculus are as follows:

$$\sin^2\theta + \cos^2\theta = 1$$

$$\tan^2\theta + 1 = \sec^2\theta$$

$$1 + \cot^2\theta = \csc^2\theta$$

$$\sin(-\theta) = -\sin\theta$$

$$\cos(-\theta) = \cos\theta$$

$$\tan(-\theta) = -\tan\theta$$

$$\sin(\theta + 2\pi) = \sin\theta$$

$$\cos(\theta + 2\pi) = \cos\theta$$

$$\tan(\theta + \pi) = \tan\theta$$

$$\sin(A + B) = \sin A\cos B + \cos A\sin B$$

$$\sin(A - B) = \sin A\cos B - \cos A\sin B$$

$$\cos(A + B) = \cos A\cos B - \sin A\sin B$$

$$\cos(A - B) = \cos A\cos B + \sin A\sin B$$

$$\tan(A + B) = \frac{\tan A + \tan B}{1 - \tan A\tan B}$$

$$\tan(A - B) = \frac{\tan A - \tan B}{1 + \tan A \tan B}$$

$$\sin 2\theta = 2\sin\theta\cos\theta$$

$$\cos 2\theta = \cos^2\theta - \sin^2\theta$$

$$= 2\cos^2\theta - 1$$

$$= 1 - 2\sin^2\theta$$

$$\sin^2 \frac{1}{2}\theta = \frac{1 - \cos\theta}{2}$$

$$\cos^2 \frac{1}{2}\theta = \frac{1 + \cos\theta}{2}$$

The relationship between the angles and sides of a triangle may be expressed using the Law of Sines or the Law of Cosines (Figure 2).

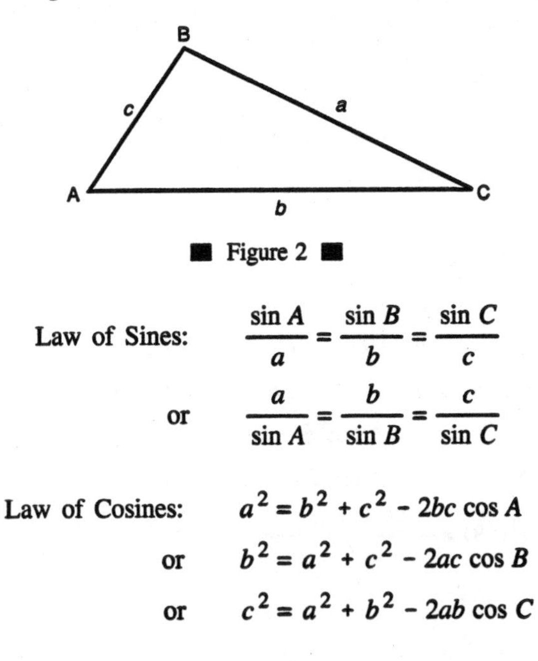


■ Figure 2 ■

Law of Sines: $\dfrac{\sin A}{a} = \dfrac{\sin B}{b} = \dfrac{\sin C}{c}$

or $\dfrac{a}{\sin A} = \dfrac{b}{\sin B} = \dfrac{c}{\sin C}$

Law of Cosines: $a^2 = b^2 + c^2 - 2bc \cos A$

or $b^2 = a^2 + c^2 - 2ac \cos B$

or $c^2 = a^2 + b^2 - 2ab \cos C$

| Degree Measure of x | Radian Measure of x | sin x | cos x | tan x |
|---|---|---|---|---|
| 0 | 0 | 0 | 1 | 0 |
| 30 | $\frac{\pi}{6}$ | $\frac{1}{2}$ | $\frac{\sqrt{3}}{2}$ | $\frac{\sqrt{3}}{3}$ |
| 45 | $\frac{\pi}{4}$ | $\frac{\sqrt{2}}{2}$ | $\frac{\sqrt{2}}{2}$ | 1 |
| 60 | $\frac{\pi}{3}$ | $\frac{\sqrt{3}}{2}$ | $\frac{1}{2}$ | $\sqrt{3}$ |
| 90 | $\frac{\pi}{2}$ | 1 | 0 | Undefined |
| 120 | $\frac{2\pi}{3}$ | $\frac{\sqrt{3}}{2}$ | $-\frac{1}{2}$ | $-\sqrt{3}$ |
| 135 | $\frac{3\pi}{4}$ | $\frac{\sqrt{2}}{2}$ | $-\frac{\sqrt{2}}{2}$ | -1 |
| 150 | $\frac{5\pi}{6}$ | $\frac{1}{2}$ | $-\frac{\sqrt{3}}{2}$ | $-\frac{\sqrt{3}}{3}$ |
| 180 | $\pi$ | 0 | -1 | 0 |
| 210 | $\frac{7\pi}{6}$ | $-\frac{1}{2}$ | $-\frac{\sqrt{3}}{2}$ | $\frac{\sqrt{3}}{3}$ |
| 225 | $\frac{5\pi}{4}$ | $-\frac{\sqrt{2}}{2}$ | $-\frac{\sqrt{2}}{2}$ | 1 |
| 240 | $\frac{4\pi}{3}$ | $-\frac{\sqrt{3}}{2}$ | $-\frac{1}{2}$ | $\sqrt{3}$ |
| 270 | $\frac{3\pi}{2}$ | -1 | 0 | Undefined |
| 300 | $\frac{5\pi}{3}$ | $-\frac{\sqrt{3}}{2}$ | $\frac{1}{2}$ | $-\sqrt{3}$ |
| 315 | $\frac{7\pi}{4}$ | $-\frac{\sqrt{2}}{2}$ | $\frac{\sqrt{2}}{2}$ | -1 |
| 330 | $\frac{11\pi}{6}$ | $-\frac{1}{2}$ | $\frac{\sqrt{3}}{2}$ | $-\frac{\sqrt{3}}{3}$ |
| 360 | $2\pi$ | 0 | 1 | 0 |

■ Table 1 ■

The concept of the limit of a function is essential to the study of calculus. It is used in defining some of the most important concepts in calculus—continuity, the derivative of a function, and the definite integral of a function.

## Intuitive Definition

The limit of a function $f(x)$ describes the behavior of the function close to a particular $x$ value. It does not necessarily give the value of the function at $x$. You write $\lim_{x \to c} f(x) = L$, which means that as $x$ "approaches" $c$, the function $f(x)$ "approaches" the real number $L$ (Figure 3).

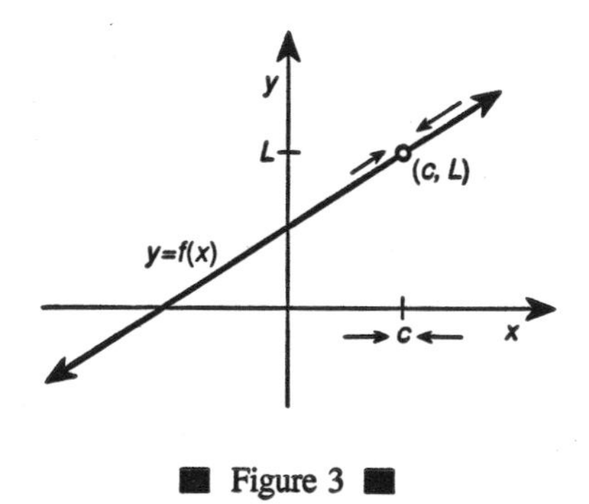


■ Figure 3 ■

In other words, as the independent variable $x$ gets closer and closer to $c$, the function value $f(x)$ gets closer to $L$. Note that this does not imply that $f(c) = L$; in fact, the function may not even exist at $c$ (Figure 4) or may equal some value different than $L$ at $c$ (Figure 5).

If the function does not approach a real number $L$ as $x$ approaches $c$, the limit does not exist; therefore, you write $\lim_{x \to c} f(x)$ DNE (Does

Not Exist). Many different situations could occur in determining that the limit of a function does not exist as $x$ approaches some value.

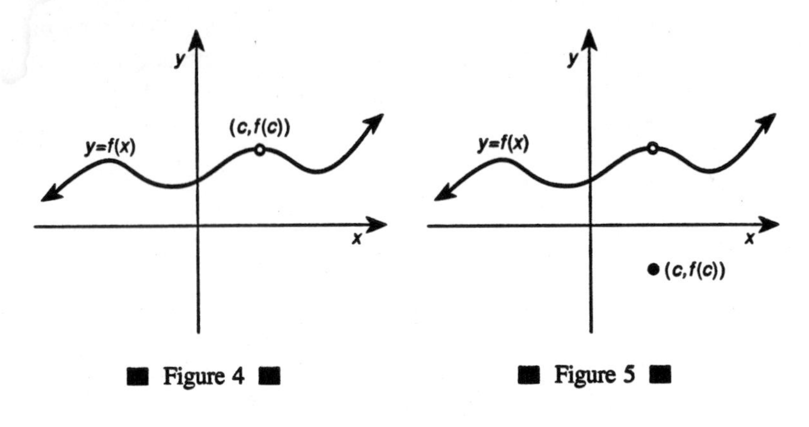


■ Figure 4 ■ ■ Figure 5 ■

## Evaluating Limits

Limits of functions are evaluated using many different techniques such as recognizing a pattern, simple substitution, or using algebraic simplifications. Some of these techniques are illustrated in the following examples.

**Example 1:** Find the limit of the sequence: $\frac{1}{2}, \frac{2}{3}, \frac{3}{4}, \frac{4}{5}, \frac{5}{6}, \frac{6}{7}, \ldots$

Because the value of each fraction gets slightly larger for each term, while the numerator is always one less than the denominator, the fraction values will get closer and closer to 1; hence, the limit of the sequence is 1.

**Example 2:** Evaluate $\lim_{x \to 2} (3x - 1)$.

When $x$ is replaced by 2, $3x$ approaches 6, and $3x - 1$ approaches 5; hence, $\lim_{x \to 2} (3x - 1) = 5$.

**Example 3:** Evaluate $\lim\limits_{x \to -3} \dfrac{x^2 - 9}{x + 3}$.

Substituting $-3$ for $x$ yields 0/0, which is meaningless. Factoring first and simplifying, you find that

$$\lim_{x \to -3} \frac{x^2 - 9}{x + 3} = \lim_{x \to -3} \frac{(x + 3)(x - 3)}{x + 3}$$

$$= \lim_{x \to -3} (x - 3)$$

$$= -6$$

The graph of $(x^2 - 9)/(x + 3)$ would be the same as the graph of the linear function $y = x - 3$ with the single point $(-3, -6)$ removed from the graph (Figure 6).

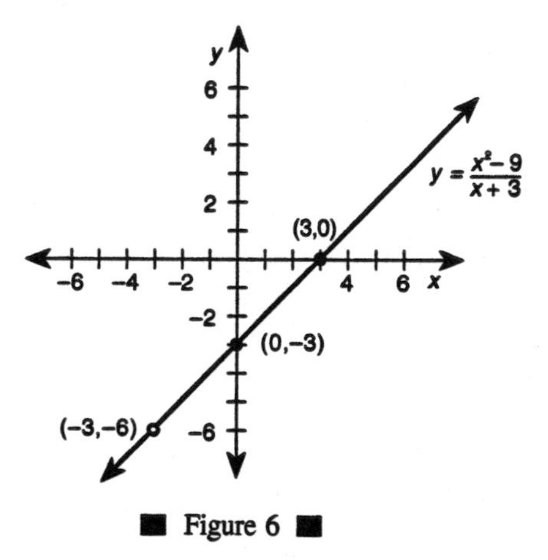


■ Figure 6 ■

**Example 4:** Evaluate $\lim_{x \to 3} \dfrac{\dfrac{x}{x+2} - \dfrac{3}{5}}{x - 3}$.

Substituting 3 for $x$ yields 0/0, which is meaningless. Simplifying the complex fraction, you find that

$$\begin{aligned}
\lim_{x \to 3} \frac{\dfrac{x}{x+2} - \dfrac{3}{5}}{x - 3} &= \lim_{x \to 3} \frac{\dfrac{x}{x+2} - \dfrac{3}{5}}{x - 3} \cdot \frac{5(x+2)}{5(x+2)} \\
&= \lim_{x \to 3} \frac{5x - 3(x+2)}{5(x+2)(x-3)} \\
&= \lim_{x \to 3} \frac{2x - 6}{5(x+2)(x-3)} \\
&= \lim_{x \to 3} \frac{2(x-3)}{5(x+2)(x-3)} \\
&= \lim_{x \to 3} \frac{2}{5(x+2)} \\
&= \frac{2}{25}
\end{aligned}$$

**Example 5:** Evaluate $\lim_{x \to 0} \dfrac{x}{x+5}$.

Substituting 0 for $x$ yields $0/5 = 0$; hence, $\lim_{x \to 0} x/(x+5) = 0$.

**Example 6:** Evaluate $\lim_{x \to 0} \dfrac{x+5}{x}$.

Substituting 0 for $x$ yields 5/0, which is meaningless; hence, $\lim_{x \to 0} (x + 5)/x$ DNE. (Remember, infinity is not a real number.)

## One-sided Limits

For some functions, it is appropriate to look at their behavior from one side only. If $x$ approaches $c$ from the right only, you write

$$\lim_{x \to c^+} f(x)$$

or if $x$ approaches $c$ from the left only, you write

$$\lim_{x \to c^-} f(x)$$

It follows, then, that $\lim_{x \to c} f(x) = L$

if and only if $\lim_{x \to c^+} f(x) = \lim_{x \to c^-} f(x) = L$

**Example 7:** Evaluate $\lim_{x \to 0^+} \sqrt{x}$.

Because $x$ is approaching 0 from the right, it is always positive; $\sqrt{x}$ is getting closer and closer to zero, so $\lim_{x \to 0^+} \sqrt{x} = 0$. Although substituting 0 for $x$ would yield the same answer, the next example illustrates why this technique is not always appropriate.

**Example 8:** Evaluate $\lim_{x \to 0^-} \sqrt{x}$.

Because $x$ is approaching 0 from the left, it is always negative, and $\sqrt{x}$ does not exist. In this situation, $\lim_{x \to 0^-} \sqrt{x}$ DNE. Also, note that $\lim_{x \to 0} \sqrt{x}$ DNE because $\lim_{x \to 0^+} \sqrt{x} = 0 \neq \lim_{x \to 0^-} \sqrt{x}$.

**Example 9:** Evaluate (a) $\lim_{x \to 2^-} \frac{|x - 2|}{x - 2}$

(b) $\lim_{x \to 2^+} \frac{|x - 2|}{x - 2}$

(c) $\lim_{x \to 2} \frac{|x - 2|}{x - 2}$

(a) As $x$ approaches 2 from the left, $x - 2$ is negative, and $|x - 2| = -(x - 2)$; hence,

$$\lim_{x \to 2^-} \frac{|x - 2|}{x - 2} = \frac{-(x - 2)}{x - 2} = -1$$

(b) As $x$ approaches 2 from the right, $x - 2$ is positive, and $|x - 2| = x - 2$; hence,

$$\lim_{x \to 2^-} \frac{|x - 2|}{x - 2} = \frac{(x - 2)}{x - 2} = 1$$

(c) Because $\lim_{x \to 2^-} \frac{|x - 2|}{x - 2} \neq \lim_{x \to 2^+} \frac{|x - 2|}{x - 2}$, $\lim_{x \to 2} \frac{|x - 2|}{x - 2}$ DNE

## Infinite Limits

Some functions "take off" in the positive or negative direction (increase or decrease without bound) near certain values for the independent variable. When this occurs, the function is said to have an infinite limit; you write $\lim_{x \to c} f(x) = +\infty$ or $\lim_{x \to c} f(x) = -\infty$. Note also that the function has a vertical asymptote at $x = c$ if either of the above limits holds true.

In general, a fractional function will have an infinite limit if the limit of the denominator is zero and the limit of the numerator is not

zero. The sign of the infinite limit is determined by the sign of the quotient of the numerator and denominator at values close to the number that the independent variable is approaching.

**Example 10:** Evaluate $\lim\limits_{x\to 0} \dfrac{1}{x^2}$.

As $x$ approaches 0, the numerator is always positive and the denominator approaches 0 and is always positive; hence, the function increases without bound and $\lim_{x\to 0} 1/x^2 = +\infty$. The function has a vertical asymptote at $x = 0$ (Figure 7).

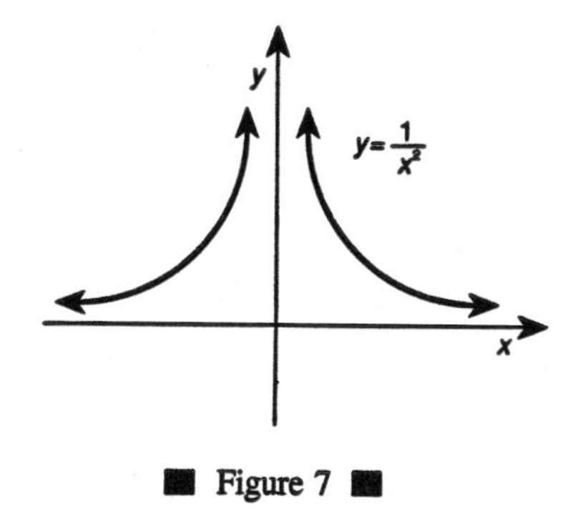


■ Figure 7 ■

**Example 11:** Evaluate $\lim\limits_{x\to 2^-} \dfrac{x+3}{x-2}$.

As $x$ approaches 2 from the left, the numerator approaches 5, and the denominator approaches 0 through negative values; hence, the function decreases without bound and $\lim_{x\to 2^-} (x + 3)/(x - 2) = -\infty$. The function has a vertical asymptote at $x = 2$.

**Example 12:** Evaluate $\lim\limits_{x\to 0^+} \left(\dfrac{1}{x^2} - \dfrac{1}{x^3}\right)$

Rewriting $1/x^2 - 1/x^3$ as an equivalent fractional expression $(x - 1)/x^3$, the numerator approaches $-1$, and the denominator approaches 0 through positive values as $x$ approaches 0 from the right; hence, the function decreases without bound and $\lim_{x\to 0^+} (1/x^2 - 1/x^3) = -\infty$. The function has a vertical asymptote at $x = 0$. A word of caution: Do not evaluate the limits individually and subtract because $\pm\infty$ are not real numbers. Using this example,

$$\lim_{x\to 0^+}\left(\frac{1}{x^2} - \frac{1}{x^2}\right) \neq \lim_{x\to 0^+}\frac{1}{x^2} - \lim_{x\to 0^+}\frac{1}{x^3} = (+\infty) - (+\infty)$$

which is meaningless.

## Limits at Infinity

Limits at infinity are used to describe the behavior of functions as the independent variable increases or decreases without bound. If a function approaches a numerical value $L$ in either of these situations, write

$$\lim_{x\to +\infty} f(x) = L \quad \text{or} \quad \lim_{x\to -\infty} f(x) = L$$

and $f(x)$ is said to have a horizontal asymptote at $y = L$. A function may have different horizontal asymptotes in each direction, have a horizontal asymptote in one direction only, or have no horizontal asymptotes.

**Example 13:** Evaluate $\lim_{x\to +\infty} \dfrac{2x^2 + 3}{x^2 - 5x - 1}$.

Factor the largest power of $x$ in the numerator from each term and the largest power of $x$ in the denominator from each term.

You find that

$$\lim_{x \to +\infty} \frac{2x^2 + 3}{x^2 - 5x - 1} = \lim_{x \to +\infty} \frac{x^2\left(2 + \frac{3}{x^2}\right)}{x^2\left(1 - \frac{5}{x} - \frac{1}{x^2}\right)}$$

$$= \lim_{x \to +\infty} \frac{2 + \frac{3}{x^2}}{1 - \frac{5}{x} - \frac{1}{x^2}}$$

$$= \frac{2 + 0}{1 - 0 - 0}$$

$$\lim_{x \to +\infty} \frac{2x^2 + 3}{x^2 - 5x - 1} = 2$$

The function has a horizontal asymptote at $y = 2$.

**Example 14:** Evaluate $\lim_{x \to +\infty} \frac{x^3 - 2}{5x^4 - 3x^3 + 2x}$.

Factor $x^3$ from each term in the numerator and $x^4$ from each term in the denominator, which yields

$$\lim_{x \to -\infty} \frac{x^3 - 2}{5x^4 - 3x^3 + 2x} = \lim_{x \to -\infty} \frac{x^3\left(1 - \frac{2}{x^3}\right)}{x^4\left(5 - \frac{3}{x} + \frac{2}{x^3}\right)}$$

$$= \lim_{x \to -\infty} \left(\frac{1}{x}\right)\left(\frac{1 - \frac{2}{x^3}}{5 - \frac{3}{x} + \frac{2}{x^3}}\right)$$

$$= (0)\left(\frac{1 - 0}{5 - 0 + 0}\right)$$

$$= 0$$

The function has a horizontal asymptote at $y = 0$.

**Example 15:** Evaluate $\lim_{x \to +\infty} \frac{9x^2}{x + 2}$.

Factor $x^2$ from each term in the numerator and $x$ from each term in the denominator, which yields

$$\lim_{x \to +\infty} \frac{9x^2}{x + 2} = \lim_{x \to +\infty} \frac{x^2(9)}{x\left(1 + \frac{2}{x}\right)}$$

$$= \lim_{x \to +\infty} x\left(\frac{9}{1 + \frac{2}{x}}\right)$$

$$= \left[\lim_{x \to +\infty} (x)\right]\left[\frac{9}{1 + 0}\right]$$

$$= [\lim_{x \to +\infty} (x)]\ [9]$$

$$\lim_{x \to +\infty} \frac{9x^2}{x+2} = +\infty$$

Because this limit does not approach a real number value, the function has no horizontal asymptote as $x$ increases without bound.

**Example 16:** Evaluate $\lim_{x \to -\infty} (x^3 - x^2 - 3x)$.

Factor $x^3$ from each term of the expression, which yields

$$\lim_{x \to -\infty} (x^3 - x^2 - 3x) = \lim_{x \to -\infty} (x^3)\left(1 - \frac{1}{x} - \frac{3}{x^2}\right)$$

$$= \lim_{x \to -\infty} (x^3) \cdot \lim_{x \to -\infty} \left(1 - \frac{1}{x} - \frac{3}{x^2}\right)$$

$$= \lim_{x \to -\infty} (x^3) \cdot [1 - 0 - 0]$$

$$= [\lim_{x \to -\infty} (x^3)] \cdot [1]$$

$$\lim_{x \to -\infty} (x^3 - x^2 - 3x) = -\infty$$

As in the previous example, this function has no horizontal asymptote as $x$ decreases without bound.

## Limits Involving Trigonometric Functions

The trigonometric functions sine and cosine have four important limit properties:

$$\lim_{x \to c} \sin x = \sin c$$

$$\lim_{x \to c} \cos x = \cos c$$

$$\lim_{x \to 0} \frac{\sin x}{x} = 1$$

and

$$\lim_{x \to 0} \frac{1 - \cos x}{x} = 0$$

These properties can be used to evaluate many limit problems involving the six basic trigonometric functions.

**Example 17:** Evaluate $\lim_{x \to 0} \frac{\cos x}{\sin x - 3}$.

Substituting 0 for $x$, you find that $\cos x$ approaches 1 and $\sin x - 3$ approaches $-3$; hence,

$$\lim_{x \to 0} \frac{\cos x}{\sin x - 3} = -\frac{1}{3}$$

**Example 18:** Evaluate $\lim_{x \to 0^+} \cot x$.

Because $\cot x = \cos x / \sin x$, you find $\lim_{x \to 0^+} \cos x / \sin x$. The numerator approaches 1 and the denominator approaches 0 through positive values because we are approaching 0 in the first quadrant; hence, the function increases without bound and $\lim_{x \to 0^+} \cot x = +\infty$, and the function has a vertical asymptote at $x = 0$.

**Example 19:** Evaluate $\lim_{x\to 0} \dfrac{\sin 4x}{x}$.

Multiplying the numerator and denominator by 4 produces

$$\lim_{x\to 0} \frac{\sin 4x}{x} = \lim_{x\to 0} \frac{4\sin 4x}{4x}$$

$$= \left(\lim_{x\to 0} 4\right) \cdot \lim_{x\to 0} \frac{\sin 4x}{4x}$$

$$= 4 \cdot 1$$

$$\lim_{x\to 0} \frac{\sin 4x}{x} = 4$$

**Example 20:** Evaluate $\lim_{x\to 0} \dfrac{\sec x - 1}{x}$.

Because $\sec x = 1/\cos x$, you find that

$$\lim_{x\to 0} \frac{\sec x - 1}{x} = \lim_{x\to 0} \frac{\frac{1}{\cos x} - 1}{x}$$

$$= \lim_{x\to 0} \frac{1 - \cos x}{x \cos x}$$

$$= \lim_{x\to 0} \left(\frac{1}{\cos x}\right) \cdot \left(\frac{1 - \cos x}{x}\right)$$

$$= \left[\lim_{x\to 0} \frac{1}{\cos x}\right] \cdot \left[\lim_{x\to 0} \frac{1 - \cos x}{x}\right]$$

$$= 1 \cdot 0$$

$$\lim_{x \to 0} \frac{\sec x - 1}{x} = 0$$

## Continuity

A function $f(x)$ is said to be continuous at a point $(c, f(c))$ if each of the following conditions is satisfied:

(1) $f(c)$ exists ($c$ is in the domain of $f$),
(2) $\lim_{x \to c} f(x)$ exists, and
(3) $\lim_{x \to c} f(x) = f(c)$.

Geometrically, this means that there is no gap, split, or missing point for $f(x)$ at $c$ and that a pencil could be moved along the graph of $f(x)$ through $(c, f(c))$ without lifting it off the graph. A function is said to be continuous at $(c, f(c))$ from the right if $\lim_{x \to c^+} f(x) = f(c)$ and continuous at $(c, f(c))$ from the left if $\lim_{x \to c^-} f(x) = f(c)$). Many of our familiar functions such as linear, quadratic and other polynomial functions, rational functions, and the trigonometric functions are continuous at each point in their domain.

A special function that is often used to illustrate one-sided limits is the greatest integer function. The greatest integer function, $[x]$, is defined to be the largest integer less than or equal to $x$ (Figure 8).

Some values of $[x]$ for specific $x$ values are

$$[2] = 2$$

$$[5.8] = 5$$

$$\left[-3\frac{1}{3}\right] = -4$$

$$[.46] = 0$$

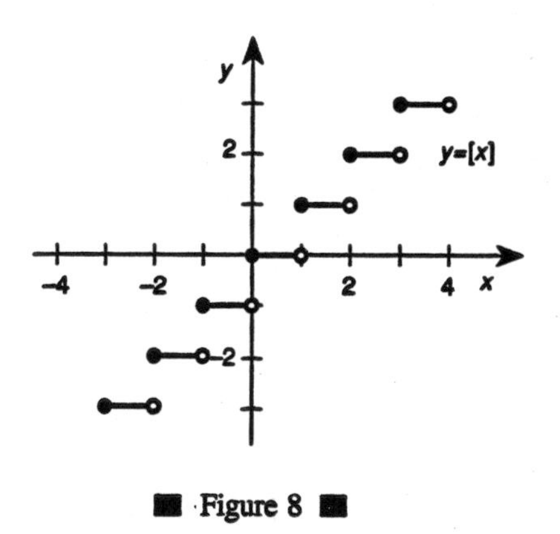


■ Figure 8 ■

The greatest integer function is continuous at any integer $n$ from the right only because

$$f(n) = [n] = n$$

and
$$\lim_{x \to n^+} f(x) = n$$

but
$$\lim_{x \to n^-} f(x) = n - 1$$

hence, $\lim_{x \to n^-} f(x) \neq f(n)$ and $f(x)$ is not continuous at $n$ from the left. Note that the greatest integer function is continuous from the right and from the left at any noninteger value of $x$.

**Example 21:** Discuss the continuity of $f(x) = 2x + 3$ at $x = -4$.

When the definition of continuity is applied to $f(x)$ at $x = -4$, you find that

(1) $f(-4) = -5$

(2) $\lim_{x \to -4} f(x) = \lim_{x \to -4} (2x + 3) = -5$

(3) $\lim_{x \to -4} f(x) = f(-4)$

hence, $f$ is continuous at $x = -4$.

**Example 22:** Discuss the continuity of $f(x) = \dfrac{x^2 - 4}{x - 2}$ at $x = 2$.

When the definition of continuity is applied to $f(x)$ at $x = 2$, you find that $f(2)$ does not exist; hence, $f$ is not continuous (discontinuous) at $x = 2$.

**Example 23:** Discuss the continuity of $f(x) = \begin{cases} \dfrac{x^2 - 4}{x - 2}, & x \neq 2 \\ 4, & x = 2 \end{cases}$ at $x = 2$.

When the definition of continuity is applied to $f(x)$ at $x = 2$, you find that

(1) $f(2) = 4$

(2)
$$\lim_{x \to 2} f(x) = \lim_{x \to 2} \frac{x^2 - 4}{x - 2}$$
$$= \lim_{x \to 2} \frac{(x - 2)(x + 2)}{x - 2}$$
$$= \lim_{x \to 2} (x + 2)$$
$$\lim_{x \to 2} f(x) = 4$$

(3) $\lim_{x \to 2} f(x) = f(2)$

hence, $f$ is continuous at $x = 2$.

**Example 24:** Discuss the continuity of $f(x) = \sqrt{x}$ at $x = 0$.

When the definition of continuity is applied to $f(x)$ at $x = 0$, you find that

(1) $f(0) = 0$

(2) $\lim_{x \to 0} f(x) = \lim_{x \to 0} \sqrt{x}$ DNE because $\lim_{x \to 0^+} \sqrt{x} = 0$,

but $\lim_{x \to 0^-} \sqrt{x}$ DNE

(3) $\lim_{x \to 0^+} f(x) = f(0)$

hence, $f$ is continuous at $x = 0$ from the right only.

**Example 25:** Discuss the continuity of $f(x) = \begin{cases} 5 - 2x, & x < -3 \\ x^2 + 2, & x \geq -3 \end{cases}$ at $x = -3$.

When the definition of continuity is applied to $f(x)$ at $x = -3$, you find that

(1) $f(-3) = (-3)^2 + 2 = 11$

(2) $\lim_{x \to -3^-} f(x) = \lim_{x \to -3^-} (5 - 2x) = 11$

$\lim_{x \to -3^+} f(x) = \lim_{x \to -3^+} (x^2 + 2) = 11$

hence, $\lim_{x \to 3} f(x) = 11$ because $\lim_{x \to -3^-} f(x) = \lim_{x \to -3^+} f(x)$

(3) $\lim_{x \to -3} f(x) = f(-3)$

hence, $f$ is continuous at $x = -3$.

Many theorems in calculus require that functions be continuous on intervals of real numbers. A function $f(x)$ is said to be continuous on an open interval $(a,b)$ if $f$ is continuous at each point $c \in (a,b)$. A function $f(x)$ is said to be continuous on a closed interval $[a,b]$ if $f$ is continuous at each point $c \in (a,b)$ and if $f$ is continuous at $a$ from the right and continuous at $b$ from the left.

**Example 26:**

(a) $f(x) = 2x + 3$ is continuous on $(-\infty,+\infty)$ because $f$ is continuous at every point $c \in (-\infty,+\infty)$.

(b) $f(x) = (x - 3)/(x + 4)$ is continuous on $(-\infty,-4)$ and $(-4,+\infty)$ because $f$ is continuous at every point $c \in (-\infty,-4)$ and $c \in (-4,+\infty)$.

(c) $f(x) = (x - 3)/(x + 4)$ is not continuous on $(-\infty,-4]$ or $[-4,+\infty)$ because $f$ is not continuous at $-4$ from the left or from the right.

(d) $f(x) = \sqrt{x}$ is continuous on $[0,+\infty)$ because $f$ is continuous at every point $c \in (0,+\infty)$ and is continuous at 0 from the right.

(e) $f(x) = \cos x$ is continuous on $(-\infty,+\infty)$ because $f$ is continuous at every point $c \in (-\infty,+\infty)$.

(f) $f(x) = \tan x$ is continuous on $(0,\pi/2)$ because $f$ is continuous at every point $c \in (0,\pi/2)$.

(g) $f(x) = \tan x$ is not continuous on $[0,\pi/2]$ because $f$ is not continuous at $\pi/2$ from the left.

(h) $f(x) = \tan x$ is continuous on $[0,\pi/2)$ because $f$ is continuous at every point $c \in (0,\pi/2)$ and is continuous at 0 from the right.

(i) $f(x) = 2x/(x^2 + 5)$ is continuous on $(-\infty,+\infty)$ because $f$ is continuous at every point $c \in (-\infty,+\infty)$.

(j) $f(x) = |x - 2|/(x - 2)$ is continuous on $(-\infty,2)$ and $(2,+\infty)$ because $f$ is continuous at every point $c \in (-\infty,2)$ and $c \in (2,+\infty)$.

(k) $f(x) = |x - 2|/(x - 2)$ is not continuous on $(-\infty,2]$ or $[2,+\infty)$ because $f$ is not continuous at 2 from the left or from the right.

One of the most important applications of limits is the concept of the derivative of a function. In calculus, the derivative is used in a wide variety of problems, and understanding it is essential to applying it to such problems.

## Definition

The **derivative** of a function $y = f(x)$ at a point $(x, f(x))$ is defined as

$$\lim_{\Delta x \to 0} \frac{f(x + \Delta x) - f(x)}{\Delta x}$$

if this limit exists. The derivative is denoted by $f'(x)$, read "$f$ prime of $x$" or "$f$ prime at $x$," and $f$ is said to be **differentiable** at $x$ if this limit exists (Figure 9).

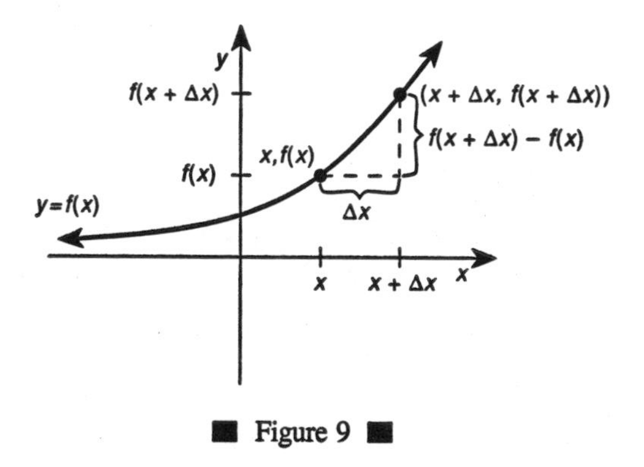


■ Figure 9 ■

If a function is differentiable at $x$, then it must be continuous at $x$, but the converse is not necessarily true. That is, a function may be continuous at a point, but the derivative at that point may not exist. As an example, the function $f(x) = x^{1/3}$ is continuous over its entire domain of real numbers, but its derivative does not exist at $x = 0$.

Another example is the function $f(x) = |x + 2|$, which is also continuous over its entire domain of real numbers but is not differentiable at $x = -2$. The relationship between continuity and differentiability can be summarized as follows: Differentiability implies continuity, but continuity *does not* imply differentiability.

**Example 1:** Find the derivative of $f(x) = x^2 - 5$ at the point $(2,-1)$.

$$\frac{f(x + \Delta x) - f(x)}{\Delta x} = \frac{(x + \Delta x)^2 - 5 - x^2 + 5}{\Delta x}$$

$$= \frac{x^2 + 2x\Delta x + \Delta x^2 - 5 - x^2 + 5}{\Delta x}$$

$$= \frac{2x\Delta x + \Delta x^2}{\Delta x}$$

$$= \frac{\Delta x(2x + \Delta x)}{\Delta x}$$

$$\frac{f(x + \Delta x) - f(x)}{\Delta x} = 2x + \Delta x$$

$$f'(x) = \lim_{\Delta x \to 0} (2x + \Delta x) = 2x$$

$$f'(2) = 2 \cdot 2 = 4$$

hence, the derivative of $f(x) = x^2 - 5$ at the point $(2,-1)$ is 4.

One interpretation of the derivative of a function at a point is the **slope of the tangent line** at this point. The derivative may be thought of as the limit of the slopes of the secant lines passing through a fixed point on a curve and other points on the curve that get closer and closer to the fixed point. If this limit exists, it is defined to be the slope of the tangent line at the fixed point $(x, f(x))$ on the graph of $y = f(x)$.

Another interpretation of the derivative is the **instantaneous velocity** of a function representing the position of a particle along a line at time $t$, where $y = s(t)$. The derivative may be thought of as the limit of the average velocities between a fixed time and other times that get closer and closer to the fixed time. If this limit exists, it is defined to be the instantaneous velocity at time $t$ for the function, $y = s(t)$.

A third interpretation of the derivative is the **instantaneous rate of change** of a function at a point. The derivative may be thought of as the limit of the average rates of change between a fixed point and other points on the curve that get closer and closer to the fixed point. If this limit exists, it is defined to be the instantaneous rate of change at the fixed point $(x, f(x))$ on the graph of $y = f(x)$.

**Example 2:** Find the instantaneous velocity of $s(t) = \dfrac{1}{t+2}$ at the time $t = 3$.

$$\frac{s(t+\Delta t) - s(t)}{\Delta t} = \frac{\dfrac{1}{t+\Delta t+2} - \dfrac{1}{t+2}}{\Delta t}$$

$$= \frac{\dfrac{1}{t+\Delta t+2} - \dfrac{1}{t+2}}{\Delta t} \cdot \frac{(t+2)(t+\Delta t+2)}{(t+2)(t+\Delta t+2)}$$

$$= \frac{(t+2) - (t+\Delta t+2)}{\Delta t(t+2)(t+\Delta t+2)}$$

$$= \frac{-\Delta t}{\Delta t(t+2)(t+\Delta t+2)}$$

$$\frac{s(t+\Delta t) - s(t)}{\Delta t} = \frac{-1}{(t+2)(t+\Delta t+2)}$$

$$s'(t) = \lim_{\Delta t \to 0} \frac{-1}{(t+2)(t+\Delta t+2)}$$

$$= \frac{-1}{(t+2)^2}$$

$$s'(3) = \frac{-1}{(3+2)^2} = \frac{-1}{5^2} = \frac{-1}{25}$$

hence, the instantaneous velocity of $s(t) = 1/(t+2)$ at time $t = 3$ is $-1/25$. The negative velocity indicates that the particle is moving in the negative direction.

A number of different notations are used to represent the derivative of a function $y = f(x)$ with $f'(x)$ being the most common. Some others are $y'$, $dy/dx$, $df/dx$, $df(x)/dx$, $D_x f$, and $D_x f(x)$, and you should be able to use any of these in selected problems.

## Differentiation Rules

Many differentiation rules can be proven using the limit definition of the derivative and are also useful in finding the derivatives of applicable functions. To eliminate the need of using the formal definition for every application of the derivative, some of the more useful formulas are listed here.

(1) If $f(x) = c$, where $c$ is a constant, then $f'(x) = 0$.

(2) If $f(x) = c \cdot g(x)$, then $f'(x) = c \cdot g'(x)$.

(3) Sum Rule: If $f(x) = g(x) + h(x)$, then $f'(x) = g'(x) + h'(x)$.

(4) Difference Rule: If $f(x) = g(x) - h(x)$, then $f'(x) = g'(x) - h'(x)$.

(5) Product Rule: If $f(x) = g(x) \cdot h(x)$, then $f'(x) = g'(x) \cdot h(x) + h'(x) \cdot g(x)$.

(6) Quotient Rule: If $f(x) = \frac{g(x)}{h(x)}$ and $h(x) \neq 0$, then

$$f'(x) = \frac{g'(x) \cdot h(x) - h'(x) \cdot g(x)}{[h(x)]^2}.$$

(7) Power Rule: If $f(x) = x^n$, then $f'(x) = nx^{n-1}$.

**Example 3:** Find $f'(x)$ if $f(x) = 6x^3 + 5x^2 + 9$.

$$\begin{aligned} f'(x) &= 6 \cdot 3x^2 + 5 \cdot 2x^1 + 0 \\ &= 18x^2 + 10x \end{aligned}$$

**Example 4:** Find $y'$ if $y = (3x + 4)(2x^2 - 3x + 5)$.

$$\begin{aligned} y' &= (3)(2x^2 - 3x + 5) + (4x - 3)(3x + 4) \\ &= 6x^2 - 9x + 15 + 12x^2 + 7x - 12 \\ &= 18x^2 - 2x + 3 \end{aligned}$$

**Example 5:** Find $\frac{dy}{dx}$ if $y = \frac{3x + 5}{2x - 3}$.

$$\begin{aligned} \frac{dy}{dx} &= \frac{3(2x - 3) - 2(3x + 5)}{(2x - 3)^2} \\ &= \frac{6x - 9 - 6x - 10}{(2x - 3)^2} \\ &= \frac{-19}{(2x - 3)^2} \end{aligned}$$

**Example 6:** Find $f'(x)$ if $f(x) = x^5 - \sqrt{x} + \frac{1}{x^3}$.

Because
$$f(x) = x^5 - x^{1/2} + x^{-3}$$
$$f'(x) = 5x^4 - \frac{1}{2}x^{-1/2} - 3x^{-4}$$
$$= 5x^4 - \frac{1}{2\sqrt{x}} - \frac{3}{x^4}$$

**Example 7:** Find $f'(3)$ if $f(x) = x^2 - 8x + 3$.

$$f'(x) = 2x - 8$$
$$f'(3) = (2)(3) - 8$$
$$= -2$$

**Example 8:** If $y = \frac{4}{x + 2}$, find $y'$ at (2,1).

$$y' = \frac{0(x + 2) - 1(4)}{(x + 2)^2}$$
$$= \frac{-4}{(x + 2)^2}$$

At (2,1),
$$y' = \frac{-4}{(2 + 2)^2}$$
$$= \frac{-4}{16}$$
$$= -\frac{1}{4}$$

**Example 9:** Find the slope of the tangent line to the curve $y = 12 - 3x^2$ at the point $(-1,9)$.

Because the slope of the tangent line to a curve is the derivative, you find that $y' = -6x$; hence, at $(-1,9)$, $y' = 6$, and the tangent line has slope 6 at the point $(-1,9)$.

## Trigonometric Function Differentiation

The six trigonometric functions also have differentiation formulas that can be used in application problems of the derivative. The rules are summarized as follows:

(1) If $f(x) = \sin x$, then $f'(x) = \cos x$.
(2) If $f(x) = \cos x$, then $f'(x) = -\sin x$.
(3) If $f(x) = \tan x$, then $f'(x) = \sec^2 x$.
(4) If $f(x) = \cot x$, then $f'(x) = -\csc^2 x$.
(5) If $f(x) = \sec x$, then $f'(x) = \sec x \tan x$.
(6) If $f(x) = \csc x$, then $f'(x) = -\csc x \cot x$.

Note that rules (3) to (6) can be proven using the quotient rule along with the given function expressed in terms of the sine and cosine functions, as illustrated in the following example.

**Example 10:** Use the definition of the tangent function and the quotient rule to prove if $f(x) = \tan x$, then $f'(x) = \sec^2 x$.

$$f(x) = \tan x$$
$$= \frac{\sin x}{\cos x}$$
$$f'(x) = \frac{\cos x \cdot \cos x - (-\sin x) \sin x}{\cos^2 x}$$

$$= \frac{\cos^2 x + \sin^2 x}{\cos^2 x}$$

$$= \frac{1}{\cos^2 x}$$

$$= \sec^2 x$$

**Example 11:** Find $y'$ if $y = x^3 \cot x$.

$$y' = 3x^2 \cot x + x^3(-\csc^2 x)$$

$$= 3x^2 \cot x - x^3 \csc^2 x$$

**Example 12:** Find $f'\left(\frac{\pi}{4}\right)$ if $f(x) = 5 \sin x + \cos x$.

$$f'(x) = 5 \cos x - \sin x$$

$$f'\left(\frac{\pi}{4}\right) = 5 \cos \frac{\pi}{4} - \sin \frac{\pi}{4}$$

$$= \frac{5\sqrt{2}}{2} - \frac{\sqrt{2}}{2}$$

$$= \frac{4\sqrt{2}}{2}$$

$$= 2\sqrt{2}$$

**Example 13:** Find the slope of the tangent line to the curve $y = \sin x$ at the point $(\pi/2,1)$.

Because the slope of the tangent line to a curve is the derivative, you find that $y' = \cos x$; hence, at $(\pi/2,1)$, $y' = \cos \pi/2 = 0$ and the tangent line has slope 0 at the point $(\pi/2,1)$. Note that the geometric interpretation of this result is that the tangent line is horizontal at this point on the graph of $y = \sin x$.

## Chain Rule

The **chain rule** provides us a technique for finding the derivative of composite functions, with the number of functions that make up the composition determining how many differentiation steps are necessary. For example, if a composite function $f(x)$ is defined as

$$f(x) = (g \circ h)(x) = g[h(x)]$$

then $$f'(x) = g'[h(x)] \cdot h'(x)$$

Note that because two functions, $g$ and $h$, make up the composite function $f$, you have to consider the derivatives $g'$ and $h'$ in differentiating $f(x)$.

If a composite function $r(x)$ is defined as

$$r(x) = (m \circ n \circ p)(x) = m\{n[p(x)]\}$$

then $$r'(x) = m'\{n[p(x)]\} \cdot n'[p(x)] \cdot p'(x)$$

Here, three functions—$m$, $n$, and $p$—make up the composition function $r$; hence, you have to consider the derivatives $m'$, $n'$, and $p'$ in differentiating $r(x)$. A technique that is sometimes suggested for differentiating composite functions is to work from the "outside to the inside" functions to establish a sequence for each of the derivatives that must be taken.

**Example 14:** Find $f'(x)$ if $f(x) = (3x^2 + 5x - 2)^8$.

$$f'(x) = 8(3x^2 + 5x - 2)^7 \cdot (6x + 5)$$
$$= 8(6x + 5)(3x^2 + 5x - 2)^7$$

Note that the chain rule allows us to find the derivative of $f(x)$ without having to first expand the polynomial to the eighth power.

**Example 15:** Find $f'(x)$ if $f(x) = \tan(\sec x)$.

$$f'(x) = \sec^2(\sec x) \cdot \sec x \tan x$$
$$= \sec x \tan x \sec^2(\sec x)$$

**Example 16:** Find $\frac{dy}{dx}$ if $y = \sin^3(3x - 1)$.

$$\frac{dy}{dx} = 3\sin^2(3x - 1) \cdot \cos(3x - 1) \cdot (3)$$
$$= 9\cos(3x - 1)\sin^2(3x - 1)$$

**Example 17:** Find $f'(2)$ if $f(x) = \sqrt{5x^2 + 3x - 1}$.

Because
$$f(x) = (5x^2 + 3x - 1)^{1/2}$$
$$f'(x) = \frac{1}{2}(5x^2 + 3x - 1)^{-1/2}(10x + 3)$$
$$= \frac{10x + 3}{2\sqrt{5x^2 + 3x - 1}}$$

$$f'(2) = \frac{10 \cdot 2 + 3}{2\sqrt{5(2)^2 + 3 \cdot 2 - 1}}$$

$$= \frac{23}{2\sqrt{25}}$$

$$= \frac{23}{10}$$

**Example 18:** Find the slope of the tangent line to the curve $y = (x^2 - 3)^5$ at the point $(-1,-32)$.

Because the slope of the tangent line to a curve is the derivative, you find that

$$y' = 5(x^2 - 3)^4(2x)$$

$$= 10x(x^2 - 3)^4$$

hence, at $(-1,-32)$ $\quad y' = 10(-1)[(-1)^2 - 3]^4$

$$= (-10)(-2)^4$$

$$= -160$$

which represents the slope of the tangent line at the point $(-1,-32)$.

## Implicit Differentiation

In mathematics, some equations in $x$ and $y$ do not explicitly define $y$ as a function of $x$ and cannot be easily manipulated to solve for $y$ in terms of $x$, even though such a function may exist. When this occurs, it is implied that there exists a function $y = f(x)$ such that the given equation is satisfied. The technique of **implicit differentiation** allows us to find the derivative of $y$ with respect to $x$ without having to solve the given equation for $y$. The chain rule must be used whenever the function $y$ is being differentiated because of our assumption that $y$ may be expressed as a function of $x$.

**Example 19:** Find $\frac{dy}{dx}$ if $x^2y^3 - xy = 10$.

Differentiating implicitly with respect to $x$, you find that

$$2xy^3 + x^2 \cdot 3y^2 \cdot \frac{dy}{dx} - 1y - x \cdot 1 \cdot \frac{dy}{dx} = 0$$

$$3x^2y^2 \frac{dy}{dx} - x\frac{dy}{dx} = y - 2xy^3$$

$$(3x^2y^2 - x)\frac{dy}{dx} = y - 2xy^3$$

$$\frac{dy}{dx} = \frac{y - 2xy^3}{3x^2y^2 - x}$$

or

$$\frac{dy}{dx} = \frac{2xy^3 - y}{x - 3x^2y^2}$$

**Example 20:** Find $y'$ if $y = \sin x + \cos y$.

Differentiating implicitly with respect to $x$, you find that

$$1 \cdot y' = \cos x - \sin y \cdot y'$$

$$1 \cdot y' + \sin y \cdot y' = \cos x$$

$$y'(1 + \sin y) = \cos x$$

$$y' = \frac{\cos x}{1 + \sin y}$$

**Example 21:** Find $y'$ at $(-1,1)$ if $x^2 + 3xy + y^2 = -1$.

Differentiating implicitly with respect to $x$, you find that

$$2x + 3y + 3x \cdot y' + 2y \cdot y' = 0$$
$$3x \cdot y' + 2y \cdot y' = -2x - 3y$$
$$y'(3x + 2y) = -2x - 3y$$
$$y' = \frac{-2x - 3y}{3x + 2y}$$

At the point $(-1,1)$,

$$y' = \frac{(-2)(-1) - 3(1)}{3(-1) + 2(1)}$$
$$= \frac{-1}{-1}$$
$$= 1$$

**Example 22:** Find the slope of the tangent line to the curve $x^2 + y^2 = 25$ at the point $(3,-4)$.

Because the slope of the tangent line to a curve is the derivative, differentiate implicitly with respect to $x$, which yields

$$2x + 2y \cdot y' = 0$$
$$2y \cdot y' = -2x$$
$$y' = \frac{-2x}{2y}$$
$$= \frac{-x}{y}$$

hence, at $(3,-4)$, $y' = -3/-4 = 3/4$, and the tangent line has slope 3/4 at the point $(3,-4)$.

## Higher Order Derivatives

Because the derivative of a function $y = f(x)$ is itself a function $y' = f'(x)$, you can take the derivative of $f'(x)$, which is generally referred to as the second derivative of $f(x)$ and written $f''(x)$ or $f^{(2)}(x)$. This differentiation process can be continued to find the third, fourth, and successive derivatives of $f(x)$, which are called **higher order derivatives** of $f(x)$. Because the "prime" notation for derivatives would eventually become somewhat messy, it is preferable to use the numerical notation $f^{(n)}(x) = y^{(n)}$ to denote the $n$th derivative of $f(x)$. A later unit will provide some applications of the second derivative in curve sketching and in distance, velocity, and acceleration problems.

**Example 23:** Find the first, second, and third derivatives of $f(x) = 5x^4 - 3x^3 + 7x^2 - 9x + 2$..

$$f'(x) = 20x^3 - 9x^2 + 14x - 9$$

$$f''(x) = f^{(2)}(x) = 60x^2 - 18x + 14$$

$$f'''(x) = f^{(3)}(x) = 120x - 18$$

**Example 24:** Find the first, second, and third derivatives of $y = \sin^2 x$.

$$\begin{aligned} y' &= 2\sin x \cos x \\ y'' &= 2\cos x \cos x + 2\sin x(-\sin x) \\ &= 2\cos^2 x - 2\sin^2 x \\ y''' &= 2 \cdot 2\cos x(-\sin x) - 2 \cdot 2\sin x \cos x \\ &= -4\sin x \cos x - 4\sin x \cos x \\ &= -8\sin x \cos x \end{aligned}$$

**Example 25:** Find $f^{(3)}(4)$ if $f(x) = \sqrt{x}$.

Because $f(x) = \sqrt{x} = x^{1/2}$

$$f'(x) = \frac{1}{2}x^{-1/2}$$

$$f''(x) = -\frac{1}{4}x^{-3/2}$$

$$f'''(x) = \frac{3}{8}x^{-5/2}$$

hence, $f'''(4) = \frac{3}{8}(4)^{-5/2}$

$$= \frac{3}{8}\left(\frac{1}{32}\right)$$

$$= \frac{3}{256}$$

## Inverse Trigonometric Function Differentiation

Each of the six basic trigonometric functions have corresponding inverse functions when appropriate restrictions are placed on the domain of the original functions. All of the inverse trigonometric functions have derivatives, which are summarized as follows:

(1) If $f(x) = \sin^{-1}x = \arcsin x$, $-\frac{\pi}{2} \leq f(x) \leq \frac{\pi}{2}$ then

$$f'(x) = \frac{1}{\sqrt{1 - x^2}}.$$

(2) If $f(x) = \cos^{-1}x = \arccos x$, $0 \leq f(x) \leq \pi$ then

$$f'(x) = \frac{-1}{\sqrt{1 - x^2}}.$$

(3) If $f(x) = \tan^{-1}x = \arctan x$, $-\frac{\pi}{2} < f(x) < \frac{\pi}{2}$ then

$$f'(x) = \frac{1}{1 + x^2}.$$

(4) If $f(x) = \cot^{-1}x = \text{arccot}\, x$, $0 < f(x) < \pi$ then

$$f'(x) = \frac{-1}{1 + x^2}.$$

(5) If $f(x) = \sec^{-1}x = \text{arcsec}\, x$, $0 \leq f(x) \leq \pi$, $f(x) \neq \frac{\pi}{2}$ then

$$f'(x) = \frac{1}{x\sqrt{x^2 - 1}}.$$

(6) If $f(x) = \csc^{-1}x = \text{arccsc}\, x$, $-\frac{\pi}{2} \leq f(x) \leq \frac{\pi}{2}$, $f(x) \neq 0$

then $f'(x) = \dfrac{-1}{x\sqrt{x^2 - 1}}$.

**Example 26:** Find $f'(x)$ if $f(x) = \cos^{-1}(5x)$.

$$f'(x) = \frac{-1}{\sqrt{1 - (5x)^2}} \cdot 5$$

$$f'(x) = \frac{-5}{\sqrt{1 - 25x^2}}$$

**Example 27:** Find $y'$ if $y = \arctan\left(\sqrt{x^3}\right)$.

Because

$$y = \arctan(x^{3/2})$$

$$y' = \frac{1}{1 + (x^{3/2})^2} \cdot \frac{3}{2}x^{1/2}$$

$$= \frac{1}{1 + x^3} \cdot \frac{3}{2}x^{1/2}$$

$$y' = \frac{3\sqrt{x}}{2(1 + x^3)}$$

## Exponential and Logarithmic Differentiation

Exponential functions and their corresponding inverse functions, called logarithmic functions, have the following differentiation formulas:

(1) If $f(x) = e^x$, then $f'(x) = e^x$.

(2) If $f(x) = a^x$, $a > 0$, $a \neq 1$, then $f'(x) = (\ln a) \cdot a^x$.

(3) If $f(x) = \ln x$, then $f'(x) = \frac{1}{x}$.

(4) If $f(x) = \log_a x$, $a > 0$, $a \neq 1$, then $f'(x) = \frac{1}{(\ln a) \cdot x}$.

Note that the exponential function $f(x) = e^x$ has the special property that its derivative is the function itself, $f'(x) = e^x = f(x)$.

**Example 28:** Find $f'(x)$ if $f(x) = e^{x^2+5}$.

$$f'(x) = e^{x^2+5} \cdot 2x$$

$$f'(x) = 2x \cdot e^{x^2+5}$$

**Example 29:** Find $y'$ if $y = 5^{\sqrt{x}}$.

$$y' = (\ln 5) \cdot 5^{\sqrt{x}} \cdot \frac{1}{2}x^{-1/2}$$

$$= (\ln 5) \cdot 5^{\sqrt{x}} \cdot \frac{1}{2\sqrt{x}}$$

$$y' = \frac{(\ln 5) \cdot 5^{\sqrt{x}}}{2\sqrt{x}}$$

**Example 30:** Find $f'(x)$ if $f(x) = \ln(\sin x)$.

$$f'(x) = \frac{1}{\sin x} \cdot \cos x$$

$$= \frac{\cos x}{\sin x}$$

$$f'(x) = \cot x$$

**Example 31:** Find $\frac{dy}{dx}$ if $y = \log_{10}(4x^2 - 3x - 5)$.

$$\frac{dy}{dx} = \frac{1}{(\ln 10)(4x^2 - 3x - 5)} \cdot (8x - 3)$$

$$\frac{dy}{dx} = \frac{8x - 3}{(\ln 10)(4x^2 - 3x - 5)}$$

The derivative of a function has many applications to problems in calculus. It may be used in curve sketching; solving maximum and minimum problems; solving distance, velocity, and acceleration problems; solving related rate problems; and approximating function values.

## Tangent and Normal Lines

As previously noted, the derivative of a function at a point is the slope of the tangent line at this point. The **normal line** is defined as the line that is perpendicular to the tangent line at the point of tangency. Because the slopes of perpendicular lines (neither of which is vertical) are negative reciprocals of one another, the slope of the normal line to the graph of $f(x)$ is $-1/f'(x)$.

**Example 1:** Find the equation of the tangent line to the graph of $f(x) = \sqrt{x^2 + 3}$ at the point $(-1,2)$.

$$f(x) = (x^2 + 3)^{1/2}$$

$$f'(x) = \frac{1}{2}(x^2 + 3)^{-1/2} \cdot (2x)$$

$$f'(x) = \frac{x}{\sqrt{x^2 + 3}}$$

At the point $(-1,2)$, $f'(-1) = -1/2$ and the equation of the line is

$$y - y_1 = m(x - x_1)$$

$$y - 2 = -\frac{1}{2}(x + 1)$$

$$2y - 4 = -x - 1$$

$$x + 2y = 3$$

**Example 2:** Find the equation of the normal line to the graph of

$$f(x) = \sqrt{x^2 + 3}$$

at the point $(-1,2)$.

From Example 1, you find that $f'(-1) = -1/2$ and the slope of the normal line is $-1/f'(-1) = 2$; hence, the equation of the normal line at the point $(-1,2)$ is

$$y - y_1 = m(x - x_1)$$

$$y - 2 = 2(x + 1)$$

$$y - 2 = 2x + 2$$

$$2x - y = -4$$

## Critical Points

Points on the graph of a function where the derivative is zero or the derivative does not exist are important to consider in many application problems of the derivative. The point $(x, f(x))$ is called a **critical point** of $f(x)$ if $x$ is in the domain of the function and either $f'(x) = 0$ or $f'(x)$ does not exist. The geometric interpretation of what is taking place at a critical point is that the tangent line is either horizontal, vertical, or does not exist at that point on the curve.

**Example 3:** Find all critical points of $f(x) = x^4 - 8x^2$.

Because $f(x)$ is a polynomial function, its domain is all real numbers.

$$f'(x) = 4x^3 - 16x$$
$$f'(x) = 0 \rightarrow 4x^3 - 16x = 0$$
$$4x(x^2 - 4) = 0$$
$$4x(x + 2)(x - 2) = 0$$
$$x = 0,\ x = -2,\ x = 2$$
$$f(-2) = (-2)^4 - 8(-2)^2 = -16$$
$$f(0) = (0)^4 - 8(0)^2 = 0$$
$$f(2) = (2)^4 - 8(2)^2 = -16$$

hence, the critical points of $f(x)$ are $(-2,-16)$, $(0,0)$, and $(2,-16)$.

**Example 4:** Find all critical points of $f(x) = \sin x + \cos x$ on $[0,2\pi]$.

The domain of $f(x)$ is restricted to the closed interval $[0,2\pi]$.

$$f'(x) = \cos x - \sin x$$
$$f'(x) = 0 \rightarrow \cos x - \sin x = 0$$
$$\cos x = \sin x$$
$$x = \frac{\pi}{4}, \frac{5\pi}{4}$$

$$f\left(\frac{\pi}{4}\right) = \sin\frac{\pi}{4} + \cos\frac{\pi}{4} = \frac{\sqrt{2}}{2} + \frac{\sqrt{2}}{2} = \sqrt{2}$$

$$f\left(\frac{5\pi}{4}\right) = \sin\frac{5\pi}{4} + \cos\frac{5\pi}{4} = \frac{-\sqrt{2}}{2} + \frac{-\sqrt{2}}{2} = -\sqrt{2}$$

hence, the critical points of $f(x)$ are $(\pi/4, \sqrt{2})$ and $(5\pi/4, -\sqrt{2})$.

## Extreme Value Theorem

An important application of critical points is in determining possible maximum and minimum values of a function on certain intervals. The **Extreme Value Theorem** guarantees both a maximum and minimum value for a function under certain conditions. It states the following:

If a function $f(x)$ is continuous on a closed interval $[a,b]$, then $f(x)$ has both a maximum and minimum value on $[a,b]$.

The procedure for applying the Extreme Value Theorem is to first establish that the function is continuous on the closed interval. The next step is to determine all critical points in the given interval and evaluate the function at these critical points and at the endpoints of the interval. The largest function value from the previous step is the maximum value, and the smallest function value is the minimum value of the function on the given interval.

**Example 5:** Find the maximum and minimum values of $f(x) = \sin x + \cos x$ on $[0, 2\pi]$.

The function is continuous on $[0, 2\pi]$, and from Example 4, the critical points are $(\pi/4, \sqrt{2})$ and $(5\pi/4, -\sqrt{2})$. The function values at the endpoints of the interval are $f(0) = 1$ and $f(2\pi) = 1$; hence, the maximum function value of $f(x)$ is $\sqrt{2}$ at $x = \pi/4$, and the minimum function value is $-\sqrt{2}$ at $x = 5\pi/4$.

Note that for this example the maximum and minimum both occur at critical points of the function.

**Example 6:** Find the maximum and minimum values of $f(x) = x^4 - 3x^3 - 1$ on $[-2,2]$.

The function is continuous on $[-2,2]$, and its derivative is $f'(x) = 4x^3 - 9x^2$.

$$f'(x) = 0 \rightarrow 4x^3 - 9x^2 = 0$$

$$x^2(4x - 9) = 0$$

$$x = 0, \quad x = \frac{9}{4}$$

Because $x = 9/4$ is not in the interval $[-2,2]$, the only critical point occurs at $x = 0$ which is $(0,-1)$. The function values at the endpoints of the interval are $f(2) = -9$ and $f(-2) = 39$; hence, the maximum function value is 39 at $x = -2$, and the minimum function value is $-9$ at $x = 2$. Note the importance of the closed interval in determining which values to consider for critical points.

## Mean Value Theorem

The **Mean Value Theorem** establishes a relationship between the slope of a tangent line to a curve and the secant line through points on a curve at the endpoints of an interval. The theorem is stated as follows:

If a function $f(x)$ is continuous on a closed interval $[a,b]$ and differentiable on an open interval $(a,b)$, then at least one number $c \in (a,b)$ exists such that

$$f'(c) = \frac{f(b) - f(a)}{b - a} \qquad \text{(Figure 10)}$$

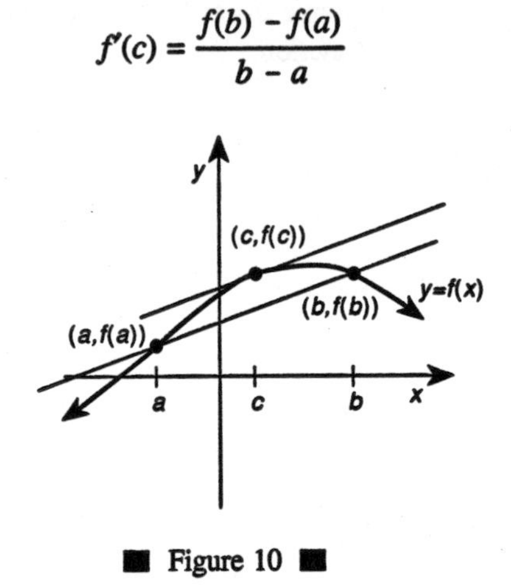


■ Figure 10 ■

Geometrically, this means that the slope of the tangent line will be equal to the slope of the secant line through $(a,f(a))$ and $(b,f(b))$ for at least one point on the curve between the two endpoints. Note that for the special case where $f(a) = f(b)$, the theorem guarantees at least one critical point, where $f'(c) = 0$ on the open interval $(a,b)$.

**Example 7:** Verify the conclusion of the Mean Value Theorem for $f(x) = x^2 - 3x - 2$ on $[-2,3]$.

The function is continuous on $[-2,3]$ and differentiable on $(-2,3)$. The slope of the secant line through the endpoint values is

$$\frac{f(3) - f(-2)}{3 - (-2)} = \frac{-2 - 8}{5} = \frac{-10}{5} = -2$$

The slope of the tangent line is

$$f'(x) = 2x - 3$$

$$f'(x) = -2 \rightarrow 2x - 3 = -2$$

$$2x = 1$$

$$x = \frac{1}{2}$$

Because $1/2 \in [-2,3]$, the $c$ value referred to in the conclusion of the Mean Value Theorem is $c = 1/2$.

## Increasing/Decreasing Functions

The derivative of a function may be used to determine whether the function is increasing or decreasing on any intervals in its domain. If $f'(x) > 0$ at each point in an interval I, then the function is said to be increasing on I. If $f'(x) < 0$ at each point in an interval I, then the function is said to be decreasing on I. Because the derivative is zero or does not exist only at critical points of the function, it must be positive or negative at all other points where the function exists.

In determining intervals where a function is increasing or decreasing, you first find domain values where all critical points will occur; then, test all intervals in the domain of the function to the left and to the right of these values to determine if the derivative is positive or negative. If $f'(x) > 0$, then $f$ is increasing on the interval, and if $f'(x) < 0$, then $f$ is decreasing on the interval. This and other information may be used to show a reasonably accurate sketch of the graph of the function.

**Example 8:** For $f(x) = x^4 - 8x^2$ determine all intervals where $f$ is increasing or decreasing.

As noted in Example 3, the domain of $f(x)$ is all real numbers, and its critical points occur at $x = -2, 0$, and 2. Testing all intervals to the left and right of these values for $f'(x) = 4x^3 - 16x$, you find that

$$f'(x) < 0 \text{ on } (-\infty, -2)$$
$$f'(x) > 0 \text{ on } (-2, 0)$$
$$f'(x) < 0 \text{ on } (0, 2)$$
$$f'(x) > 0 \text{ on } (2, +\infty)$$

hence, $f$ is increasing on $(-2,0)$ and $(2,+\infty)$ and decreasing on $(-\infty,-2)$ and $(0,2)$.

**Example 9:** For $f(x) = \sin x + \cos x$ on $[0, 2\pi]$, determine all intervals where $f$ is increasing or decreasing.

As noted in Example 4, the domain of $f(x)$ is restricted to the closed interval $[0, 2\pi]$, and its critical points occur at $\pi/4$ and $5\pi/4$. Testing all intervals to the left and right of these values for $f'(x) = \cos x - \sin x$, you find that

$$f'(x) > 0 \text{ on } \left[0, \frac{\pi}{4}\right)$$
$$f'(x) < 0 \text{ on } \left(\frac{\pi}{4}, \frac{5\pi}{4}\right)$$
$$f'(x) > 0 \text{ on } \left(\frac{5\pi}{4}, 2\pi\right]$$

hence, $f$ is increasing on $[0, \pi/4)$ and $(5\pi/4, 2\pi]$ and decreasing on $(\pi/4, 5\pi/4)$.

## First Derivative Test for Local Extrema

If the derivative of a function changes sign around a critical point, the function is said to have a local (relative) extrema at that point. If the derivative changes from positive (increasing function) to negative (decreasing function), the function has a local (relative) maximum at the critical point. If, however, the derivative changes from negative (decreasing function) to positive (increasing function), the function has a local (relative) minimum at the critical point. When this technique is used to determine local maximum or minimum function values, it is called the **First Derivative Test for Local Extrema.** Note that there is no guarantee that the derivative will change signs, and therefore, it is essential to test each interval around a critical point.

**Example 10:** If $f(x) = x^4 - 8x^2$, determine all local extrema for the function.

As noted in Example 8, $f(x)$ has critical points at $x = -2, 0, 2$. Because $f'(x)$ changes from negative to positive around $-2$ and $2$, $f$ has a local minimum at $(-2,-16)$ and $(2,-16)$. Also, $f'(x)$ changes from positive to negative around 0, and hence, $f$ has a local maximum at $(0,0)$.

**Example 11:** If $f(x) = \sin x + \cos x$ on $[0,2\pi]$, determine all local extrema for the function.

As noted in Example 9, $f(x)$ has critical points at $x = \pi/4$ and $5\pi/4$. Because $f'(x)$ changes from positive to negative around $\pi/4$, $f$ has local maximum at $(\pi/4, \sqrt{2})$. Also, $f'(x)$ changes from negative to positive around $5\pi/4$, and hence, $f$ has a local minimum at $(5\pi/4, -\sqrt{2})$.

## Second Derivative Test for Local Extrema

The second derivative may be used to determine local extrema of a function under certain conditions. If a function has a critical point for which $f'(x) = 0$ and the second derivative is positive at this point, then $f$ has a local minimum here. If, however, the function has a critical point for which $f'(x) = 0$ and the second derivative is negative at this point, then $f$ has a local maximum here. This technique is called the **Second Derivative Test for Local Extrema.**

Three possible situations could occur that would rule out the use of the Second Derivative Test for Local Extrema:

(1) $f'(x) = 0$ and $f''(x) = 0$
(2) $f'(x) = 0$ and $f''(x)$ does not exist
(3) $f'(x)$ does not exist

Under any of these conditions, the First Derivative Test would have to be used to determine any local extrema. Another drawback to the Second Derivative Test is that for some functions the second derivative is difficult or tedious to find. As with the previous situations, revert back to the First Derivative Test to determine any local extrema.

**Example 12:** Find any local extrema of $f(x) = x^4 - 8x^2$ using the Second Derivative Test.

As noted in Example 3, $f'(x) = 0$ at $x = -2, 0$, and 2. Because $f''(x) = 12x^2 - 16$, you find that $f''(-2) = 32 > 0$, and $f$ has a local minimum at $(-2, -16)$; $f''(0) = -16 < 0$, and $f$ has a local maximum at $(0,0)$; and $f''(2) = 32 > 0$, and $f$ has a local minimum at $(2, -16)$. These results agree with the local extrema determined in Example 10 using the First Derivative Test on $f(x) = x^4 - 8x^2$.

**Example 13:** Find any local extrema of $f(x) = \sin x + \cos x$ on $[0, 2\pi]$ using the Second Derivative Test.

As noted in Example 4, $f'(x) = 0$ at $x = \pi/4$ and $5\pi/4$. Because $f''(x) = -\sin x - \cos x$, you find that $f''(\pi/4) = -\sqrt{2}$ and $f$ has a local maximum at $(\pi/4, \sqrt{2})$. Also, $f''(5\pi/4) = \sqrt{2}$, and $f$ has a local minimum at $(5\pi/4, -\sqrt{2})$. These results agree with the local extrema determined in Example 11 using the First Derivative Test on $f(x) = \sin x + \cos x$ on $[0, 2\pi]$.

## Concavity and Points of Inflection

The second derivative of a function may also be used to determine the general shape of its graph on selected intervals. A function is said to be **concave upward** on an interval if $f''(x) > 0$ at each point in the interval and **concave downward** on an interval if $f''(x) < 0$ at each point in the interval. If a function changes from concave upward to concave downward or vice versa around a point, it is called a **point of inflection** of the function.

In determining intervals where a function is concave upward or concave downward, you first find domain values where $f''(x) = 0$ or $f''(x)$ does not exist. Then test all intervals around these values in the second derivative of the function. If $f''(x)$ changes sign, then $(x, f(x))$ is a point of inflection of the function. As with the First Derivative Test for Local Extrema, there is no guarantee that the second derivative will change signs, and therefore, it is essential to test each interval around the values for which $f''(x) = 0$ or does not exist.

Geometrically, a function is concave upward on an interval if its graph behaves like a portion of a parabola that opens upward. Likewise, a function that is concave downward on an interval looks like a portion of a parabola that opens downward. If the graph of a function is linear on some interval in its domain, its second derivative will be zero, and it is said to have no concavity on that interval.

**Example 14:** Determine the concavity of $f(x) = x^3 - 6x^2 - 12x + 2$ and identify any points of inflection of $f(x)$.

Because $f(x)$ is a polynomial function, its domain is all real numbers.

$$f'(x) = 3x^2 - 12x - 12$$

$$f''(x) = 6x - 12$$

$$f''(x) = 0 \rightarrow 6x - 12 = 0$$

$$6x = 12$$

$$x = 2$$

Testing the intervals to the left and right of $x = 2$ for $f''(x) = 6x - 12$, you find that

$$f''(x) < 0 \text{ on } (-\infty,2)$$

and

$$f''(x) > 0 \text{ on } (2,+\infty)$$

hence, $f$ is concave downward on $(-\infty,2)$ and concave upward on $(2,+\infty)$, and the function has a point of inflection at $(2,-38)$.

**Example 15:** Determine the concavity of $f(x) = \sin x + \cos x$ on $[0,2\pi]$ and identify any points of inflection of $f(x)$.

The domain of $f(x)$ is restricted to the closed interval $[0,2\pi]$.

$$f'(x) = \cos x - \sin x$$

$$f''(x) = -\sin x - \cos x$$

$$f''(x) = 0 \rightarrow -\sin x - \cos x = 0$$

$$-\sin x = \cos x$$

$$x = \frac{3\pi}{4}, \frac{7\pi}{4}$$

Testing all intervals to the left and right of these values for $f''(x) = -\sin x - \cos x$, you find that

$$f''(x) < 0 \text{ on } \left[0, \frac{3\pi}{4}\right]$$

$$f''(x) > 0 \text{ on } \left(\frac{3\pi}{4}, \frac{7\pi}{4}\right)$$

$$f''(x) < 0 \text{ on } \left(\frac{7\pi}{4}, 2\pi\right]$$

hence, $f$ is concave downward on $[0, 3\pi/4)$ and $(7\pi/4, 2\pi]$ and concave upward on $(3\pi/4, 7\pi/4)$ and has points of inflection at $(3\pi/4, 0)$ and $(7\pi/4, 0)$.

## Maximum/Minimum Problems

Many application problems in calculus involve functions for which you want to find maximum or minimum values. The restrictions stated or implied for such functions will determine the domain from which you must work. The function, together with its domain, will suggest which technique is appropriate to use in determining a maximum or minimum value—the Extreme Value Theorem, the First Derivative Test, or the Second Derivative Test.

**Example 16:** A rectangular box with a square base and no top is to have a volume of 108 cubic inches. Find the dimensions for the box that require the least amount of material.

The function that is to be minimized is the surface area ($S$) while the volume ($V$) remains fixed at 108 cubic inches (Figure 11).

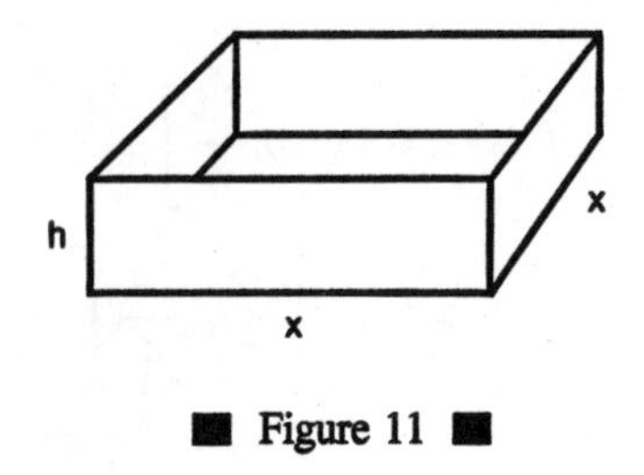


■ Figure 11 ■

Letting $x$ = length of the square base and $h$ = height of the box, you find that

$$V = x^2h = 108 \text{ cu in} \rightarrow h = \frac{108}{x^2}$$

$$S = x^2 + 4xh$$

$$S = f(x) = x^2 + 4x\left(\frac{108}{x^2}\right)$$

$$f(x) = x^2 + \frac{432}{x}$$

with the domain of $f(x) = (0,+\infty)$ because $x$ represents a length.

$$f'(x) = 2x - \frac{432}{x^2}$$

$$f'(x) = 0 \rightarrow 2x - \frac{432}{x^2} = 0$$

$$2x^3 - 432 = 0$$
$$2x^3 = 432$$
$$x^3 = 216$$
$$x = 6$$

hence, a critical point occurs when $x = 6$. Using the Second Derivative Test:

$$f''(x) = 2 + \frac{864}{x^3}$$
$$f''(6) = 6 > 0$$

and $f$ has a local minimum at $x = 6$; hence, the dimensions of the box that require the least amount of material are a length and width of 6 inches and a height of 3 inches.

**Example 17:** A right circular cylinder is inscribed in a right circular cone so that the center lines of the cylinder and the cone coincide. The cone has height 8 cm and radius 6 cm. Find the maximum volume possible for the inscribed cylinder.

The function that is to be maximized is the volume ($V$) of a cylinder inscribed in a cone with height 8 cm and radius 6 cm (Figure 12).

Letting $r$ = radius of the cylinder and $h$ = height of the cylinder and applying similar triangles, you find that

$$\frac{h}{8} = \frac{6 - r}{6}$$
$$6h = 48 - 8r$$
$$h = 8 - \frac{4}{3}r$$

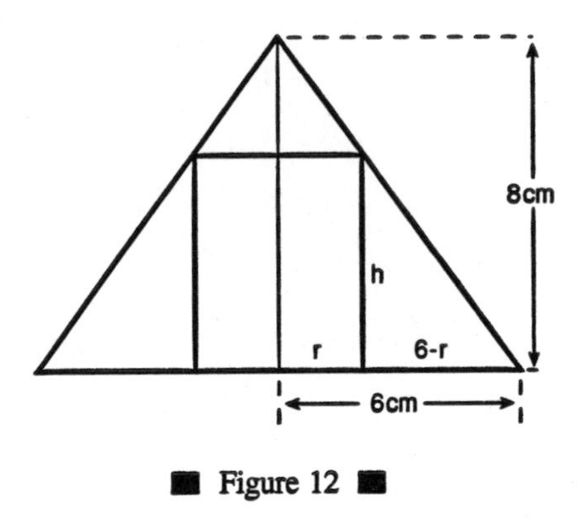


Figure 12

Because $V = \pi r^2 h$ and $h = 8 - (4/3)r$, you find that

$$V = f(r) = \pi r^2 (8 - \frac{4}{3} r)$$

$$f(r) = 8\pi r^2 - \frac{4}{3}\pi r^3$$

with the domain of $f(r) = [0,6]$ because $r$ represents the radius of the cylinder, which cannot be greater than the radius of the cone.

$$f'(r) = 16\pi r - 4\pi r^2$$

$$f'(r) = 0 \to 16\pi r - 4\pi r^2 = 0$$

$$4\pi r(4 - r) = 0$$

$$r = 0,4$$

Because $f(r)$ is continuous on [0,6], use the Extreme Value Theorem and evaluate the function at its critical points and at its endpoints; hence,

$$f(0) = 0$$

$$f(4) = \frac{128\pi}{3}$$

$$f(6) = 0$$

hence, the maximum volume is $128\pi/3$ cm$^3$, which will occur when the radius of the cylinder is 4 cm and its height is 8/3 cm.

## Distance, Velocity, and Acceleration

As previously mentioned, the derivative of a function representing the position of a particle along a line at time $t$ is the instantaneous velocity at that time. The derivative of the velocity, which is the second derivative of the position function, represents the instantaneous acceleration of the particle at time $t$.

If $y = s(t)$ represents the position function, then $v = s'(t)$ represents the instantaneous velocity, and $a = v'(t) = s''(t)$ represents the instantaneous acceleration at time t.

A positive velocity indicates that the position is increasing as time increases, while a negative velocity indicates that the position is decreasing with respect to time. If the distance remains constant, then the velocity will be zero on such an interval of time. Likewise, a positive acceleration implies that the velocity is increasing with respect to time, and a negative acceleration implies that the velocity is decreasing with respect to time. If the velocity remains constant on an interval of time, then the acceleration will be zero on the interval.

**Example 18:** The position of a particle on a line is given by $s(t) = t^3 - 3t^2 - 6t + 5$, where $t$ is measured in seconds and $s$ is measured in feet. Find

(a) The velocity of the particle at the end of 2 seconds.
(b) The acceleration of the particle at the end of 2 seconds.

Part (a): The velocity of the particle is

$$v = s'(t) = 3t^2 - 6t - 6$$

At $t = 2$ seconds $s'(2) = 3(2)^2 - 6(2) - 6$

$$s'(2) = -6 \text{ ft/sec}$$

Part (b): The acceleration of the particle is

$$a = v'(t) = s''(t) = 6t - 6$$

At $t = 2$ seconds $v'(2) = s''(2) = 6(2) - 6$

$$v'(2) = s''(2) = 6 \text{ ft/sec}^2$$

**Example 19:** The formula $s(t) = -4.9t^2 + 49t + 15$ gives the height in meters of an object after it is thrown vertically upward from a point 15 meters above the ground at a velocity of 49 m/sec. How high above the ground will the object reach?

The velocity of the object will be zero at its highest point above the ground. That is, $v = s'(t) = 0$, where

$$v = s'(t) = -9.8t + 49$$

$$s'(t) = 0 \rightarrow -9.8t + 49 = 0$$

$$-9.8t = -49$$
$$t = 5 \text{ seconds}$$

The height above the ground at 5 seconds is

$$s(5) = -4.9(5)^2 + 49(5) + 15$$
$$s(5) = 137.5 \text{ meters}$$

hence, the object will reach its highest point at 137.5 m above the ground.

## Related Rates of Change

Some problems in calculus require finding the rate of change of two or more variables that are related to a common variable, namely time. To solve these types of problems, the appropriate rate of change is determined by implicit differentiation with respect to time. Note that a given rate of change is positive if the dependent variable increases with respect to time and negative if the dependent variable decreases with respect to time. The sign of the rate of change of the solution variable with respect to time will also indicate whether the variable is increasing or decreasing with respect to time.

**Example 20:** Air is being pumped into a spherical balloon such that its radius increases at a rate of .75 in/min. Find the rate of change of its volume when the radius is 5 inches.

The volume ($V$) of a sphere with radius $r$ is

$$V = \frac{4}{3}\pi r^3$$

Differentiating with respect to $t$, you find that

$$\frac{dV}{dt} = \frac{4}{3}\pi \cdot 3r^2 \cdot \frac{dr}{dt}$$

$$\frac{dV}{dt} = 4\pi r^2 \cdot \frac{dr}{dt}$$

The rate of change of the radius $dr/dt = .75$ in/min because the radius is increasing with respect to time.

At $r = 5$ inches, you find that

$$\frac{dV}{dt} = 4\pi\,(5 \text{ inches})^2 \cdot (.75 \text{ in/min})$$

$$\frac{dV}{dt} = 75\pi \text{ cu in/min}$$

hence, the volume is increasing at a rate of $75\pi$ cu in/min when the radius has a length of 5 inches.

**Example 21:** A car is traveling north toward an intersection at a rate of 60 mph while a truck is traveling east away from the intersection at a rate of 50 mph. Find the rate of change of the distance between the car and the truck when the car is 3 miles south of the intersection and the truck is 4 miles east of the intersection.

Let $x$ = distance traveled by the truck
$y$ = distance traveled by the car
$z$ = distance between the car and truck

The distances are related by the Pythagorean Theorem: $x^2 + y^2 = z^2$ (Figure 13).

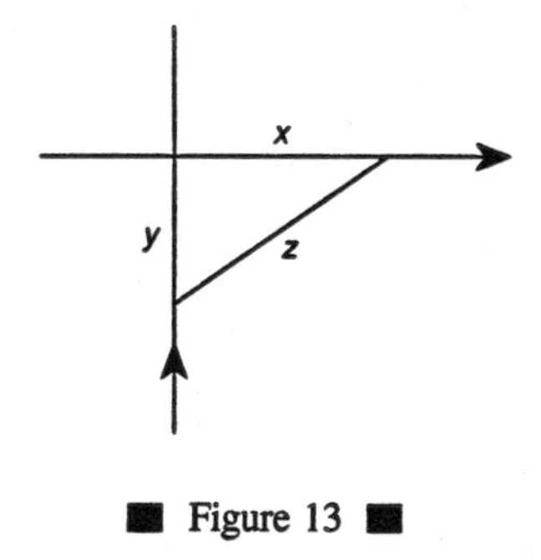


■ Figure 13 ■

The rate of change of the truck is $dx/dt = 50$ mph because it is traveling away from the intersection, while the rate of change of the car is $dy/dt = -60$ mph because it is traveling toward the intersection. Differentiating with respect to time, you find that

$$2x\frac{dx}{dt} + 2y\frac{dy}{dt} = 2z\frac{dz}{dt}$$

$$x\frac{dx}{dt} + y\frac{dy}{dt} = z\frac{dz}{dt}$$

$$(4 \text{ mi})(50 \text{ mph}) + (3 \text{ mi})(-60 \text{ mph}) = (5 \text{ mi})\frac{dz}{dt}$$

$$20 \text{ mi}^2\text{ph} = (5 \text{ mi})\frac{dz}{dt}$$

$$\frac{dz}{dt} = 4 \text{ mph}$$

hence, the distance between the car and the truck is increasing at a rate of 4 mph at the time in question.

## Differentials

The derivative of a function can often be used to approximate certain function values with a surprising degree of accuracy. To do this, the concept of the differential of the independent variable and the dependent variable must be introduced.

The definition of the derivative of a function $y = f(x)$, as you recall, is

$$f'(x) = \lim_{\Delta x \to 0} \frac{f(x + \Delta x) - f(x)}{\Delta x}$$

which represents the slope of the tangent line to the curve at some point $(x, f(x))$. If $\Delta x$ is very small ($\Delta x \neq 0$), then the slope of the tangent is approximately the same as the slope of the secant line through $(x, f(x))$. That is,

$$f'(x) \approx [f(x + \Delta x) - f(x)]/\Delta x$$

or equivalently, $$f'(x) \cdot \Delta x \approx f(x + \Delta x) - f(x)$$

The differential of the independent variable $x$ is written $dx$ and is the same as the change in $x, \Delta x$. That is,

$$dx = \Delta x,\ \Delta x \neq 0$$

hence, $$f'(x) \cdot dx \approx f(x + \Delta x) - f(x)$$

The differential of the dependent variable $y$, written $dy$, is defined to be

$$dy = f'(x) \cdot dx \approx f(x + \Delta x) - f(x)$$

Because $$\Delta y = f(x + \Delta x) - f(x)$$

you find that $$dy = f'(x)dx \approx \Delta y$$

The conclusion to be drawn from the above discussion is that the differential of $y$ ($dy$) is approximately equal to the exact change in $y$ ($\Delta y$), provided that the change in $x$ ($\Delta x = dx$) is relatively small. The smaller the change in $x$, the closer $dy$ will be to $\Delta y$, enabling you to approximate function values close to $f(x)$ (Figure 14).

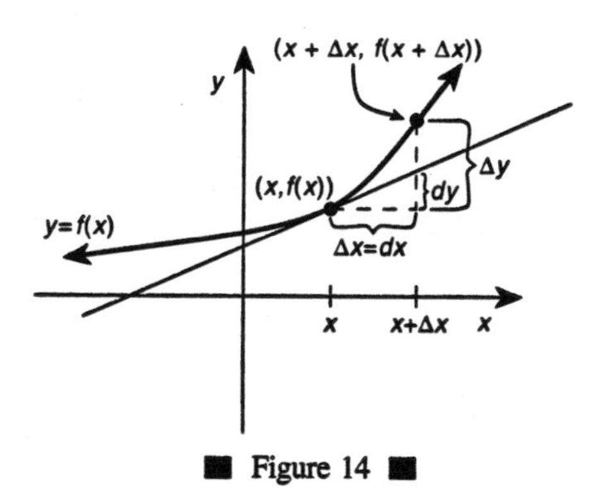


■ Figure 14 ■

**Example 22:** Find $dy$ for $y = x^3 + 5x - 1$.

Because

$$y = f(x) = x^3 + 5x - 1$$

$$f'(x) = 3x^2 + 5$$

$$dy = f'(x) \cdot dx$$

$$dy = (3x^2 + 5) \cdot dx$$

**Example 23:** Use differentials to approximate the change in the area of a square if the length of its side increases from 6 cm to 6.23 cm.

Let $x$ = length of the side of the square. The area may be expressed as a function of $x$, where $y = f(x) = x^2$. The differential $dy$ is

$$dy = f'(x) \cdot dx$$
$$dy = 2x \cdot dx$$

Because $x$ is increasing from 6 to 6.23, you find that $\Delta x = dx = .23$ cm; hence,

$$dy = 2(6 \text{ cm})(.23 \text{ cm})$$
$$dy = 2.76 \text{ cm}^2$$

The area of the square will increase by approximately 2.76 cm$^2$ as its side length increases from 6 to 6.23. Note that the exact increase in area ($\Delta y$) is 2.8129 cm$^2$.

**Example 24:** Use differentials to approximate the value of $\sqrt[3]{26.55}$ to the nearest thousandth.

Because the function you are applying is $f(x) = \sqrt[3]{x}$, choose a convenient value of $x$ that is a perfect cube and is relatively close to 26.55, namely $x = 27$. The differential $dy$ is

$$dy = f'(x)\,dx$$
$$dy = \frac{1}{3}x^{-2/3}\,dx$$
$$dy = \frac{1}{3x^{2/3}}\,dx$$

Because $x$ is decreasing from 27 to 26.55, you find that $\Delta x = dx = -.45$; hence,

$$dy = \frac{1}{3(27)^{2/3}} \cdot (-.45)$$

$$= \frac{1}{27} \cdot -\frac{45}{100}$$

$$dy = -\frac{1}{60}$$

which implies that $\sqrt[3]{26.55}$ will be approximately 1/60 less than $\sqrt[3]{27} = 3$; hence,

$$\sqrt[3]{26.55} \approx 3 - \frac{1}{60}$$

$$\approx 3 - .0167$$

$$\approx 2.9833$$

$$\sqrt[3]{26.55} \approx 2.983 \text{ to the nearest thousandth}$$

Note that the calculator value of $\sqrt[3]{26.55}$ is 2.983239874, which rounds to the same answer to the nearest thousandth!

Along with differentiation, a second important operation of calculus is antidifferentiation, or integration. These operations may be thought of as inverses of one another, and the rules for finding derivatives discussed earlier will be useful in establishing corresponding rules for finding antiderivatives. The relationship between antiderivatives and definite integrals will be discussed later with the statement of the Fundamental Theorem of Calculus (page 103).

## Antiderivatives/Indefinite Integrals

A function $F(x)$ is called an **antiderivative** of a function $f(x)$ if $F'(x) = f(x)$ for all $x$ in the domain of $f$. Note that the function $F$ is not unique and that an infinite number of antiderivatives could exist for a given function. For example, $F(x) = x^3$, $G(x) = x^3 + 5$, and $H(x) = x^3 - 2$ are all antiderivatives of $f(x) = 3x^2$ because $F'(x) = G'(x) = H'(x) = f(x)$ for all $x$ in the domain of $f$. It is clear that these functions $F$, $G$, and $H$ differ only by some constant value and that the derivative of that constant value is always zero. In other words, if $F(x)$ and $G(x)$ are antiderivatives of $f(x)$ on some interval, then $F'(x) = G'(x)$ and $F(x) = G(x) + C$ for some constant $C$ in the interval. Geometrically, this means that the graphs of $F(x)$ and $G(x)$ are identical except for their vertical position.

The notation used to represent all antiderivatives of a function $f(x)$ is the **indefinite integral** symbol written ($\int$), where $\int f(x)\,dx = F(x) + C$. The function $f(x)$ is called the integrand, and $C$ is referred to as the constant of integration. The expression $F(x) + C$ is called the indefinite integral of $f$ with respect to the independent variable $x$. Using the previous example of $F(x) = x^3$ and $f(x) = 3x^2$, you find that $\int 3x^2\,dx = x^3 + C$. The indefinite integral of a function is sometimes called the general antiderivative of the function as well.

**Example 1:** Find the indefinite integral of $f(x) = \cos x$.

Because the derivative of $F(x) = \sin x$ is $F'(x) = \cos x$, write $\int \cos x \, dx = \sin x + C$.

**Example 2:** Find the general antiderivative of $f(x) = -8$.

Because the derivative of $F(x) = -8x$ is $F'(x) = -8$, write $\int -8dx = -8x + C$.

## Integration Techniques

Many integration formulas can be derived directly from their corresponding derivative formulas, while other integration problems require more work. Some that require more work are substitution and change of variables, integration by parts, trigonometric integrals, and trigonometric substitutions.

**Basic formulas.** Most of the basic formulas given below directly follow the differentiation rules that were discussed earlier.

(1) $\int kf(x)\, dx = k \int f(x)\, dx$

(2) $\int [f(x) \pm g(x)]\, dx = \int f(x)\, dx \pm \int g(x)\, dx$

(3) $\int k\, dx = kx + C$

(4) $\int x^n\, dx = \dfrac{x^{n+1}}{n+1} + C, \ n \neq -1$

(5) $\int \sin x\, dx = -\cos x + C$

(6) $\int \cos x \, dx = \sin x + C$

(7) $\int \sec^2 x \, dx = \tan x + C$

(8) $\int \csc^2 x \, dx = -\cot x + C$

(9) $\int \sec x \tan x \, dx = \sec x + C$

(10) $\int \csc x \cot x \, dx = -\csc x + C$

(11) $\int e^x \, dx = e^x + C$

(12) $\int a^x \, dx = \dfrac{a^x}{\ln a} + C, \ a > 0, \ a \neq 1$

(13) $\int \dfrac{dx}{x} = \ln |x| + C$

(14) $\int \tan x \, dx = -\ln |\cos x| + C$

(15) $\int \cot x \, dx = \ln |\sin x| + C$

(16) $\int \sec x \, dx = \ln |\sec x + \tan x| + C$

(17) $\int \csc x \, dx = -\ln |\csc x + \cot x| + C$

(18) $\int \dfrac{dx}{\sqrt{a^2 - x^2}} = \arcsin \dfrac{x}{a} + C$

(19) $\int \dfrac{dx}{a^2 + x^2} = \dfrac{1}{a} \arctan \dfrac{x}{a} + C$

(20) $\int \dfrac{dx}{x\sqrt{x^2 - a^2}} = \dfrac{1}{a} \operatorname{arcsec} \dfrac{x}{a} + C$

**Example 3:** Evaluate $\int x^4\,dx$.

Using (4) above, you find that $\int x^4\,dx = \frac{x^5}{5} + C$.

**Example 4:** Evaluate $\int \frac{1}{\sqrt{x}}\,dx$.

Because $1/\sqrt{x} = x^{-1/2}$, using (4) above yields

$$\int \frac{1}{\sqrt{x}}\,dx = \int x^{-1/2}\,dx$$

$$= \frac{x^{1/2}}{\frac{1}{2}} + C$$

$$= 2x^{1/2} + C$$

**Example 5:** Evaluate $\int (6x^2 + 5x - 3)\,dx$.

Applying (1), (2), (3), and (4) above, you find that

$$\int (6x^2 + 5x - 3)\,dx = \frac{6x^3}{3} + \frac{5x^2}{2} - 3x + C$$

$$= 2x^3 + \frac{5}{2}x^2 - 3x + C$$

**Example 6:** Evaluate $\int \frac{dx}{x+4}$.

Using (13) above, you find that $\int \frac{dx}{x+4} = \ln|x+4| + C$.

**Example 7:** Evaluate $\int \frac{dx}{25 + x^2}$.

Using (19) above with $a = 5$, you find that

$$\int \frac{dx}{25 + x^2} = \frac{1}{5} \arctan \frac{x}{5} + C$$

**Substitution and change of variables.** One of the integration techniques that is useful in evaluating indefinite integrals that do not seem to fit the basic formulas is **substitution** and **change of variables.** This technique is often compared to the chain rule for differentiation because they both apply to composite functions. In this method, the inside function of the composition is usually replaced by a single variable (often $u$). Note that the derivative or a constant multiple of the derivative of the inside function must be a factor of the integrand.

The purpose in using the substitution technique is to rewrite the integration problem in terms of the new variable so that one or more of the basic integration formulas can then be applied. Although this approach may seem like more work initially, it will eventually make the indefinite integral much easier to evaluate. Note that for the final answer to make sense, it must be written in terms of the original variable of integration.

**Example 8:** Evaluate $\int x^2(x^3 + 1)^5\, dx$.

Because the inside function of the composition is $x^3 + 1$, substitute with

$$u = x^3 + 1$$

$$du = 3x^2\, dx$$

$$\frac{1}{3} du = x^2\, dx$$

hence,
$$\int x^2(x^3+1)^5\,dx = \frac{1}{3}\int u^5\,du$$
$$= \frac{1}{3}\cdot\frac{u^6}{6} + C$$
$$= \frac{1}{18}u^6 + C$$
$$= \frac{1}{18}(x^3+1)^6 + C$$

**Example 9:** Evaluate $\int \sin(5x)\,dx$.

Because the inside function of the composition is $5x$, substitute with

$$u = 5x$$
$$du = 5dx$$
$$\frac{1}{5}du = dx$$

hence,
$$\int \sin(5x)\,dx = \frac{1}{5}\int \sin u\,du$$
$$= -\frac{1}{5}\cos u + C$$
$$= -\frac{1}{5}\cos(5x) + C$$

**Example 10:** Evaluate $\int \frac{3x}{\sqrt{9-x^2}}\,dx$.

Because the inside function of the composition is $9 - x^2$, substitute with

$$u = 9 - x^2$$
$$du = -2x\,dx$$
$$-\frac{1}{2}du = x\,dx$$

hence,
$$\int \frac{3x}{\sqrt{9 - x^2}}\,dx = -\frac{3}{2}\int \frac{1}{\sqrt{u}}\,du$$
$$= -\frac{3}{2}\int u^{-1/2}\,du$$
$$= -\frac{3}{2} \cdot \frac{u^{1/2}}{\frac{1}{2}} + C$$
$$= -3u^{1/2} + C$$
$$= -3\sqrt{9 - x^2} + C$$

**Integration by parts.** Another integration technique to consider in evaluating indefinite integrals that do not fit the basic formulas is **integration by parts.** This method might be considered when the integrand is a single transcendental function or a product of an algebraic function and a transcendental function. The basic formula for integration by parts is

$$\int u\,dv = uv - \int v\,du$$

where $u$ and $v$ are differential functions of the variable of integration. A general rule of thumb to follow is to first choose $dv$ as the most complicated part of the integrand that can be easily integrated to find $v$. The $u$ function will be the remaining part of the integrand that will be differentiated to find $du$. The goal of this technique is to find an integral, $\int v\,du$, which is easier to evaluate than the original integral.

**Example 11:** Evaluate $\int x \sec^2 x\,dx$.

Let $u = x$ and $dv = \sec^2 x\,dx$

$du = dx \quad v = \tan x$

hence,
$$\begin{aligned}\int x \sec^2 x\,dx &= x \tan x - \int \tan x\,dx \\ &= x \tan x - (-\ln|\cos x|) + C \\ &= x \tan x + \ln|\cos x| + C\end{aligned}$$

**Example 12:** Evaluate $\int x^4 \ln x\,dx$.

Let $u = \ln x$ and $dv = x^4\,dx$

$du = \frac{1}{x}dx \quad v = \frac{x^5}{5}$

hence,
$$\begin{aligned}\int x^4 \ln x\,dx &= \frac{x^5}{5} \ln x - \int \frac{x^5}{5} \cdot \frac{1}{x}\,dx \\ &= \frac{x^5}{5} \ln x - \frac{1}{5} \int x^4\,dx \\ &= \frac{1}{5}x^5 \ln x - \frac{1}{25}x^5 + C\end{aligned}$$

**Example 13:** Evaluate $\int \arctan x\, dx$.

Let $u = \arctan x$ and $dv = dx$

$$du = \frac{1}{1+x^2}dx \qquad v = x$$

hence,

$$\int \arctan x\, dx = x \arctan x - \int \frac{x}{1+x^2}\, dx$$

$$= x \arctan x - \frac{1}{2}\ln(1+x^2) + C$$

**Trigonometric integrals.** Integrals involving powers of the trigonometric functions must often be manipulated to get them into a form in which the basic integration formulas can be applied. It is extremely important for you to be familiar with the basic trigonometric identities that were reviewed in the first section of the text because these will often be used to rewrite the integrand in a more workable form. As it was in the previous method, the goal is to find an integral that is easier to evaluate than the original integral.

**Example 14:** Evaluate $\int \cos^3 x \sin^4 x\, dx$.

$$\int \cos^3 x \sin^4 x\, dx = \int \cos^2 x \sin^4 x \cos x\, dx$$

$$= \int (1 - \sin^2 x) \sin^4 x \cos x\, dx$$

$$= \int (\sin^4 x - \sin^6 x) \cos x\, dx$$

$$= \int \sin^4 x \cos x \, dx - \int \sin^6 x \cos x \, dx$$
$$= \frac{1}{5}\sin^5 x - \frac{1}{7}\sin^7 x + C$$

**Example 15:** Evaluate $\int \sec^6 x \, dx$.

$$\begin{aligned}
\int \sec^6 x \, dx &= \int \sec^4 x \sec^2 x \, dx \\
&= \int (\sec^2 x)^2 \sec^2 x \, dx \\
&= \int (\tan^2 x + 1)^2 \sec^2 x \, dx \\
&= \int (\tan^4 x + 2\tan^2 x + 1) \sec^2 x \, dx \\
&= \int \tan^4 x \sec^2 x \, dx + \int 2\tan^2 x \sec^2 x \, dx + \int \sec^2 x \, dx \\
&= \frac{1}{5}\tan^5 x + \frac{2}{3}\tan^3 x + \tan x + C
\end{aligned}$$

**Example 16:** Evaluate $\int \sin^4 x \, dx$.

$$\begin{aligned}
\int \sin^4 x \, dx &= \int (\sin^2 x)^2 \, dx \\
&= \int \left(\frac{1 - \cos 2x}{2}\right)^2 dx \\
&= \frac{1}{4}\int (1 - 2\cos 2x + \cos^2 2x) \, dx \\
&= \frac{1}{4}\int \left(1 - 2\cos 2x + \frac{1 + \cos 4x}{2}\right) dx
\end{aligned}$$

$$= \frac{1}{4}\int\left(\frac{3}{2} - 2\cos 2x + \frac{\cos 4x}{2}\right)dx$$

$$= \frac{1}{8}\int(3 - 4\cos 2x + \cos 4x)\,dx$$

$$= \frac{1}{8}\left(3x - 2\sin 2x + \frac{1}{4}\sin 4x\right) + C$$

$$= \frac{3}{8}x - \frac{1}{4}\sin 2x + \frac{1}{32}\sin 4x + C$$

**Trigonometric substitutions.** If an integrand contains a radical expression of the form $\sqrt{a^2 - x^2}$, $\sqrt{a^2 + x^2}$, or $\sqrt{x^2 - a^2}$, a specific **trigonometric substitution** may be helpful in evaluating the indefinite integral. Some general rules to follow are

(1) If the integrand contains $\sqrt{a^2 - x^2}$

let $x = a \sin \theta$

$dx = a \cos \theta \, d\theta$

and $\sqrt{a^2 - x^2} = a \cos \theta$

(2) If the integrand contains $\sqrt{a^2 + x^2}$

let $x = a \tan \theta$

$dx = a \sec^2\theta \, d\theta$

and $\sqrt{a^2 + x^2} = a \sec \theta$

(3) If the integrand contains $\sqrt{x^2 - a^2}$

let $x = a \sec \theta$

$dx = a \sec \theta \tan \theta \, d\theta$

and $\sqrt{x^2 - a^2} = a \tan \theta$

Right triangles may be used in each of the three cases presented above to determine the expressions for any of the six trigonometric functions that appear in the evaluation of the indefinite integral.

**Example 17:** Evaluate $\int \frac{dx}{x^2\sqrt{4 - x^2}}$.

Because the radical has the form $\sqrt{a^2 - x^2}$

let $x = a \sin \theta = 2 \sin \theta$

$dx = 2 \cos \theta \, d\theta$

and $\sqrt{4 - x^2} = 2 \cos \theta$ (Figure 15)

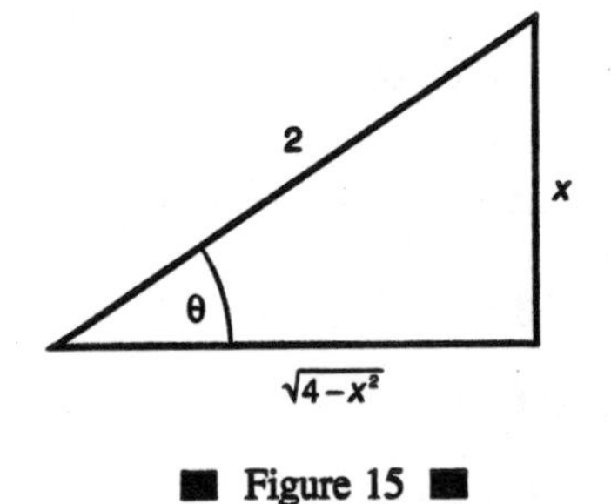


■ Figure 15 ■

hence,
$$\int \frac{dx}{x^2\sqrt{4 - x^2}} = \int \frac{2 \cos \theta \, d\theta}{(4 \sin^2 \theta)(2 \cos \theta)}$$

$$= \frac{1}{4}\int \frac{d\theta}{\sin^2\theta}$$

$$= \frac{1}{4}\int \csc^2\theta\, d\theta$$

$$= -\frac{1}{4}\cot\theta + C$$

$$= -\frac{1}{4}\cdot\frac{\sqrt{4-x^2}}{x} + C$$

$$= -\frac{\sqrt{4-x^2}}{4x} + C$$

**Example 18:** Evaluate $\int \frac{dx}{\sqrt{25+x^2}}$.

Because the radical has the form $\sqrt{a^2+x^2}$
let $x = a\tan\theta = 5\tan\theta$

$dx = 5\sec^2\theta\, d\theta$

and $\sqrt{25+x^2} = 5\sec\theta$ (Figure 16)

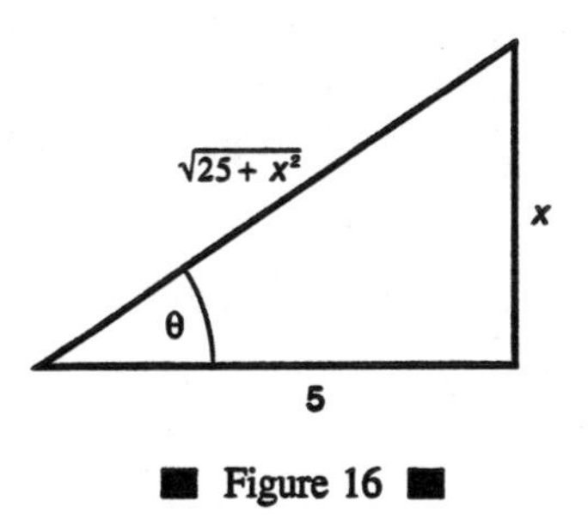


■ Figure 16 ■

hence, $$\int \frac{dx}{\sqrt{25 + x^2}} = \int \frac{5\sec^2\theta \, d\theta}{5\sec\theta}$$

$$= \int \sec\theta \, d\theta$$

$$= \ln|\sec\theta + \tan\theta| + C$$

$$= \ln\left|\frac{\sqrt{25 + x^2}}{5} + \frac{x}{5}\right| + C$$

## Distance, Velocity, and Acceleration

The indefinite integral is commonly applied in problems involving distance, velocity, and acceleration, each of which is a function of time. In the discussion of the applications of the derivative, note that the derivative of a distance function represents instantaneous velocity and that the derivative of the velocity function represents instantaneous acceleration at a particular time. In considering the relationship between the derivative and the indefinite integral as inverse operations, note that the indefinite integral of the acceleration function represents the velocity function and that the indefinite integral of the velocity represents the distance function.

In the case of a free-falling object, the acceleration due to gravity is $-32$ ft/sec$^2$. The significance of the negative is that the rate of change of the velocity with respect to time (acceleration), is negative because the velocity is decreasing as the time increases. Using the fact that the velocity is the indefinite integral of the acceleration, you find that

$$a(t) = s''(t) = -32$$

$$v(t) = s'(t) = \int s''(t) \, dt$$

$$= \int -32dt$$

$$= -32t + C_1$$

Now, at $t = 0$, the initial velocity ($v_0$) is

$$v_0 = v(0) = (-32)(0) + C_1$$

$$v_0 = C_1$$

hence, because the constant of integration for the velocity in this situation is equal to the initial velocity, write $v(t) = -32t + v_0$.

Because the distance is the indefinite integral of the velocity, you find that

$$s(t) = \int v(t)\,dt$$

$$= \int (-32t + v_0)\,dt$$

$$= -32 \cdot \frac{t^2}{2} + v_0t + C_2$$

$$= -16t^2 + v_0t + C_2$$

Now, at $t = 0$, the initial distance ($s_0$) is

$$s_0 = s(0) = -16(0)^2 + v_0(0) + C_2$$

$$s_0 = C_2$$

hence, because the constant of integration for the distance in this situation is equal to the initial distance, write $s(t) = -16t^2 + v_0(t) + s_0$.

**Example 19:** A ball is thrown downward from a height of 512 feet with a velocity of 64 feet per second. How long will it take for the ball to reach the ground?

From the given conditions, you find that

$$a(t) = -32 \text{ ft/sec}^2$$

$$v_0 = -64 \text{ ft/sec}$$

$$s_0 = 512 \text{ ft}$$

hence,

$$v(t) = -32t - 64$$

$$s(t) = -16t^2 - 64t + 512$$

The distance is zero when the ball reaches the ground or

$$-16t^2 - 64t + 512 = 0$$

$$-16(t^2 + 4t - 32) = 0$$

$$-16(t + 8)(t - 4) = 0$$

$$t = -8, t = 4$$

hence, the ball will reach the ground 4 seconds after it is thrown.

**Example 20:** In the previous example, what will the velocity of the ball be when it hits the ground?

Because $v(t) = -32(t) - 64$ and it takes 4 seconds for the ball to reach the ground, you find that

$$v(4) = -32(4) - 64$$

$$= -192 \text{ ft/sec}$$

hence, the ball will hit the ground with a velocity of $-192$ ft/sec. The significance of the negative velocity is that the rate of change of the distance with respect to time (velocity) is negative because the distance is decreasing as the time increases.

**Example 21:** A missile is accelerating at a rate of $4t$ m/sec$^2$ from a position at rest in a silo 35 m below ground level. How high above the ground will it be after 6 seconds?

From the given conditions, you find that $a(t) = 4t$ m/sec$^2$, $v_0 = 0$ m/sec because it begins at rest, and $s_0 = -35$ m because the missile is below ground level; hence,

$$v(t) = \int 4t\, dt = 2t^2$$

and

$$s(t) = \int 2t^2\, dt = \frac{2}{3}t^3 - 35$$

After 6 seconds, you find that

$$s(6) = \frac{2}{3}(6)^3 - 35$$

$$= 109 \text{ m}$$

hence, the missile will be 109 m above the ground after 6 seconds.

## Definite Integrals

The **definite integral** of a function is closely related to the antiderivative and indefinite integral of a function. The primary difference is that the definite integral, if it exists, is a real number value, while the latter two represent an infinite number of functions that differ only by a constant. The relationship between these concepts will be discussed in the section on the Fundamental Theorem of Calculus (page 103), and you will see that the definite integral will have applications to many problems in calculus.

**Definition of definite integrals.** The development of the definition of the definite integral begins with a function $f(x)$, which is continuous on a closed interval $[a,b]$. The given interval is partitioned into "$n$" subintervals that, although not necessary, can be taken to be of equal lengths ($\Delta x$). An arbitrary domain value, $x_i$, is chosen in each subinterval, and its subsequent function value, $f(x_i)$, is determined. The product of each function value times the corresponding subinterval length is determined, and these "$n$" products are added to determine their sum. This sum is referred to as a **Riemann sum** and may be positive, negative, or zero, depending upon the behavior of the function on the closed interval. For example, if $f(x) \geq 0$ on $[a,b]$, then the Riemann sum will be a positive real number. If $f(x) \leq 0$ on $[a,b]$, then the Riemann sum will be a negative real number. The Riemann sum of the function $f(x)$ on $[a,b]$ is expressed as

$$S_n = f(x_1)\Delta x + f(x_2)\Delta x + f(x_3)\Delta x + \cdots + f(x_n)\Delta x$$

or

$$S_n = \sum_{i=1}^{n} f(x_i)\,\Delta x$$

A Riemann sum may, therefore, be thought of as a "sum of $n$ products."

**Example 22:** Evaluate the Riemann sum for $f(x) = x^2$ on [1,3] using four subintervals of equal length, where $x_i$ is the right endpoint in the $i$th subinterval.

Because the subintervals are to be of equal lengths, you find that

$$\Delta x = \frac{b - a}{n}$$
$$= \frac{3 - 1}{4}$$
$$= \frac{1}{2}$$

The Riemann sum for four subintervals is

$$S_4 = \sum_{i=1}^{4} f(x_i)\,\Delta x \qquad \text{(Figure 17)}$$
$$= f(x_1)\Delta x + f(x_2)\Delta x + f(x_3)\Delta x + f(x_4)\Delta x$$
$$= [f(x_1) + f(x_2) + f(x_3) + f(x_4)]\Delta x$$
$$= \left[f\left(\frac{3}{2}\right) + f(2) + f\left(\frac{5}{2}\right) + f(3)\right] \cdot \frac{1}{2}$$
$$= \left[\frac{9}{4} + 4 + \frac{25}{4} + 9\right] \cdot \frac{1}{2}$$
$$= \left[\frac{86}{4}\right] \cdot \frac{1}{2}$$
$$S_4 = \frac{43}{4}$$

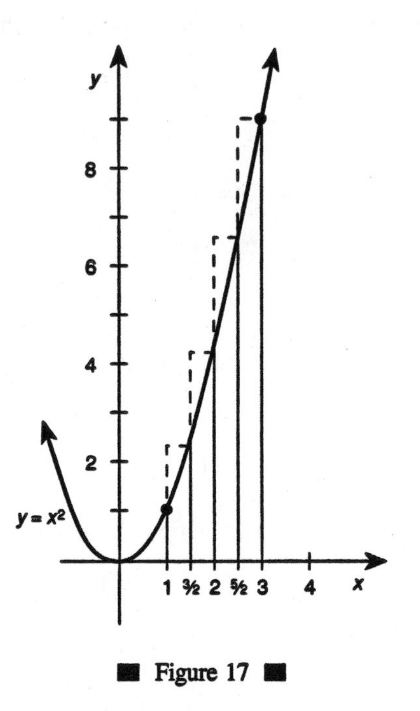


■ Figure 17 ■

If the number of subintervals is increased repeatedly, the effect would be that the length of each subinterval would get smaller and smaller. This may be restated as "if the number of subintervals increases without bound ($n \to +\infty$), then the length of each subinterval approaches zero ($\Delta x \to 0$). This limit of a Riemann sum, if it exists, is used to define the definite integral of a function on $[a,b]$. If $f(x)$ is defined on the closed interval $[a,b]$ then the **definite integral** of $f(x)$ from $a$ to $b$ is defined as

$$\int_a^b f(x)\,dx = \lim_{n \to +\infty} S_n$$

$$= \lim_{n \to +\infty} \sum_{i=1}^{n} f(x_i)\,\Delta x$$

$$= \lim_{\Delta x \to 0} \sum_{i=1}^{n} f(x_i)\,\Delta x$$

if this limit exists.

The function $f(x)$ is called the integrand, and the variable $x$ is the variable of integration. The numbers $a$ and $b$ are called the limits of integration with $a$ referred to as the lower limit of integration while $b$ is referred to as the upper limit of integration. Note that the symbol $\int$, used with the definite integral, is the same symbol used previously for the indefinite integral of a function. The reason for this will be made more apparent in the discussion of the Fundamental Theorem of Calculus (page 103). Also, keep in mind that the definite integral is a unique real number and does not represent an infinite number of functions that result from the indefinite integral of a function.

The question of the existence of the limit of a Riemann sum is important to consider because it determines whether the definite integral exists for a function on a closed interval. As with differentiation, a significant relationship exists between continuity and integration and is summarized as follows: If a function $f(x)$ is continuous on a closed interval $[a,b]$, then the definite integral of $f(x)$ on $[a,b]$ exists and $f$ is said to be integrable on $[a,b]$. If a function is not continuous on $[a,b]$, it may or may not be integrable on $[a,b]$. In other words, continuity guarantees that the definite integral exists, but the converse is not necessarily true. Unfortunately, the fact that the definite integral of a function exists on a closed interval does not imply that the value of the definite integral is easy to find.

**Properties of definite integrals.** Certain properties are useful in solving problems requiring the application of the definite integral. Some of the more common properties are

(1) $\int_a^a f(x)\,dx = 0$

(2) $\int_a^b f(x)\,dx = -\int_b^a f(x)\,dx$

(3) $\int_a^b c\,dx = c(b - a)$, where $c$ is a constant

(4) $\int_a^b cf(x)\,dx = c\int_a^b f(x)\,dx$

(5) Sum Rule: $\int_a^b [f(x) + g(x)]\,dx = \int_a^b f(x)\,dx + \int_a^b g(x)\,dx$

(6) Difference Rule: $\int_a^b [f(x) - g(x)]\,dx$

$= \int_a^b f(x)\,dx - \int_a^b g(x)\,dx$

(7) If $f(x) \geq 0$ on $[a,b]$, then $\int_a^b f(x)\,dx \geq 0$

(8) If $f(x) \leq 0$ on $[a,b]$, then $\int_a^b f(x)\,dx \leq 0$

(9) If $f(x) \geq g(x)$ on $[a,b]$, then $\int_a^b f(x)\,dx \geq \int_a^b g(x)\,dx$

(10) If $a$, $b$, and $c$ are any three points on a closed interval, then

$$\int_a^b f(x)\,dx = \int_a^c f(x)\,dx + \int_c^b f(x)\,dx$$

(11) The Mean Value Theorem for Definite Integrals: If $f(x)$ is continuous on the closed interval $[a,b]$, then at least one number $c$ exists in the open interval $(a,b)$ such that

$$\int_a^b f(x)\,dx = f(c)(b - a)$$

The value of $f(c)$ is called the average or mean value of the function $f(x)$ on the interval $[a,b]$ and

$$f(c) = \frac{1}{b - a}\int_a^b f(x)\,dx$$

**Example 23:** Evaluate $\int_2^6 3\,dx$.

$$\int_2^6 3\,dx = 3(6-2)$$
$$= 12$$

**Example 24:** Given that $\int_0^3 x^2\,dx = 9$, evaluate $\int_0^3 -4x^2\,dx$.

$$\int_0^3 -4x^2\,dx = -4\int_0^3 x^2\,dx$$
$$= (-4)\cdot 9$$
$$= -36$$

**Example 25:** Given that $\int_4^9 \sqrt{x}\,dx = \frac{38}{3}$, evaluate $\int_9^4 \sqrt{x}\,dx$.

$$\int_9^4 \sqrt{x}\,dx = -\int_4^9 \sqrt{x}\,dx$$
$$= -\frac{38}{3}$$

**Example 26:** Evaluate $\int_3^3 (x^3 + 5x^2 - 3x + 11)\,dx$.

$$\int_3^3 (x^3 + 5x^2 - 3x + 11)\,dx = 0$$

**Example 27:** Given that $\int_1^3 f(x)\,dx = 6$ and $\int_1^3 g(x)\,dx = 10$, evaluate $\int_1^3 [f(x) + g(x)]\,dx$.

$$\int_1^3 [f(x) + g(x)]\,dx = \int_1^3 f(x)\,dx + \int_1^3 g(x)\,dx$$

$$= 6 + 10$$

$$= 16$$

**Example 28:** Given that $\int_3^7 f(x)\,dx = -2$ and $\int_3^7 g(x)\,dx = 9$, evaluate $\int_3^7 [f(x) - g(x)]\,dx$.

$$\int_3^7 [f(x) - g(x)]\,dx = \int_3^7 f(x)\,dx - \int_3^7 g(x)\,dx$$

$$= -2 - 9$$

$$= -11$$

**Example 29:** Given that $\int_2^9 f(x)\,dx = 12$ and $\int_6^9 f(x) = 7$, evaluate $\int_2^6 f(x)\,dx$.

$$\int_2^9 f(x)\,dx = \int_2^6 f(x)\,dx + \int_6^9 f(x)\,dx$$

$$\int_2^6 f(x)\,dx = \int_2^9 f(x)\,dx - \int_6^9 f(x)\,dx$$

$$= 12 - 7$$

$$= 5$$

**Example 30:** Given that $\int_3^6 (x^2 - 2)\,dx = 57$

find all $c$ values that satisfy the Mean Value Theorem for the given function on the closed interval.

By the Mean Value Theorem,

$$\int_a^b f(x)\,dx = f(c)(b-a)$$

for some $c$ in $(a,b)$,

and
$$f(c) = \frac{1}{b-a}\int_a^b f(x)\,dx$$

hence,
$$f(c) = \frac{1}{b-a}\int_3^6 (x^2-2)\,dx$$
$$= \frac{1}{3}\cdot 57$$
$$= 19$$

Because
$$f(x) = x^2 - 2, \quad f(c) = c^2 - 2$$

and
$$c^2 - 2 = 19$$
$$c^2 = 21$$
$$c = \pm\sqrt{21}$$

Because $\sqrt{21} \approx 4.58$ is in the interval (3,6), the conclusion of the Mean Value Theorem is satisfied for this value of $c$.

**The Fundamental Theorem of Calculus.** The Fundamental Theorem of Calculus establishes the relationship between indefinite and definite integrals and introduces a technique for evaluating definite integrals without using Riemann sums, which is very important because evaluating the limit of a Riemann sum can be extremely time-consuming and difficult. The statement of the theorem is: If $f(x)$ is continuous on the interval $[a,b]$, and $F(x)$ is any antiderivative of $f(x)$ on $[a,b]$, then $\int_a^b f(x)\,dx = F(b) - F(a) = F(x)\Big]_a^b$.

In other words, the value of the definite integral of a function on $[a,b]$ is the difference of any antiderivative of the function evaluated at the upper limit of integration minus the same antiderivative evaluated at the lower limit of integration. Because the constants of integration are the same for both parts of this difference, they are ignored in the evaluation of the definite integral because they subtract and yield zero. Keeping this in mind, choose the constant of integration to be zero for all definite integral evaluations after Example 31.

**Example 31:** Evaluate $\int_2^5 x^2\,dx$.

Because the general antiderivative of $x^2$ is $(1/3)x^3 + C$, you find that

$$\begin{aligned}\int_2^5 x^2\,dx &= \left[\frac{1}{3}x^3 + C\right]_2^5 \\ &= \left[\frac{1}{3}(5)^3 + C\right] - \left[\frac{1}{3}(2)^3 + C\right] \\ &= \frac{125}{3} - \frac{8}{3} \\ &= 39\end{aligned}$$

**Example 32:** Evaluate $\int_{\pi/3}^{2\pi} \sin x\,dx$.

Because an antiderivative of $\sin x$ is $-\cos x$, you find that

$$\int_{\pi/3}^{2\pi} \sin x\,dx = -\cos x\Big]_{\pi/3}^{2\pi}$$

$$= (-1) - \left(-\frac{1}{2}\right)$$
$$= -\frac{1}{2}$$

**Example 33:** Evaluate $\int_1^4 \sqrt{x}\, dx$.

Because $\sqrt{x} = x^{1/2}$, an antiderivative of $x^{1/2}$ is $\frac{2}{3}x^{3/2}$, and you find that

$$\int_1^4 x^{1/2}\, dx = \frac{2}{3}x^{3/2}\Big]_1^4$$
$$= \frac{2}{3}(4)^{3/2} - \frac{2}{3}(1)^{3/2}$$
$$= \frac{16}{3} - \frac{2}{3}$$
$$= \frac{14}{3}$$

**Example 34:** Evaluate $\int_1^3 (x^2 - 4x + 1)\, dx$.

Because an antiderivative of $x^2 - 4x + 1$ is $(1/3)x^3 - 2x^2 + x$, you find that

$$\int_1^3 (x^2 - 4x + 1)\, dx = \left[\frac{1}{3}x^3 - 2x^2 + x\right]_1^3$$
$$= \left[\frac{1}{3}(3)^3 - 2(3)^2 + 3\right] - \left[\frac{1}{3}(1)^3 - 2(1)^2 + 1\right]$$

$$= (-6) - \left(-\frac{2}{3}\right)$$
$$= -\frac{16}{3}$$

**Definite integral evaluation.** The numerous techniques that can be used to evaluate indefinite integrals can also be used to evaluate definite integrals. The methods of substitution and change of variables, integration by parts, trigonometric integrals, and trigonometric substitutions are illustrated in the following examples.

**Example 35:** Evaluate $\int_1^2 \frac{x\,dx}{(x^2+2)^3}$.

Using the substitution method with

$$u = x^2 + 2$$
$$du = 2x\,dx$$
$$\frac{1}{2}du = x\,dx$$

the limits of integration can be converted from $x$ values to their corresponding $u$ values. When $x = 1$, $u = 3$ and when $x = 2$, $u = 6$, you find that

$$\int_1^2 \frac{x\,dx}{(x^2+2)^3} = \frac{1}{2}\int_2^6 \frac{du}{u^3}$$
$$= \frac{1}{2}\int_2^6 u^{-3}\,du$$

$$= \frac{1}{2}\left[-\frac{1}{2}u^{-2}\right]_2^6$$

$$= -\frac{1}{4}\left[(6)^{-2} - (2)^{-2}\right]$$

$$= -\frac{1}{4}\left(\frac{1}{36} - \frac{1}{4}\right)$$

$$= \frac{1}{18}$$

Note that when the substitution method is used to evaluate definite integrals, it is not necessary to go back to the original variable if the limits of integration are converted to the new variable values.

**Example 36:** Evaluate $\int_{\pi}^{3\pi/2} \sqrt{\sin x + 1}\, \cos x\, dx$.

Using the substitution method with $u = \sin x + 1$, $du = \cos x\, dx$, you find that $u = 1$ when $x = \pi$ and $u = 0$ when $x = 3\pi/2$; hence,

$$\int_{\pi}^{3\pi/2} \sqrt{\sin x + 1}\, \cos x\, dx = \int_1^0 u^{1/2}\, du$$

$$= \frac{2}{3}u^{3/2}\Big]_1^0$$

$$= \frac{2}{3}\left[0^{3/2} - 1^{3/2}\right]$$

$$= -\frac{2}{3}$$

Note that you never had to return to the trigonometric functions in the original integral to evaluate the definite integral.

**Example 37:** Evaluate $\int_{\pi/3}^{\pi/2} x \sin x\, dx$.

Using integration by parts with

$$u = x \text{ and } dv = \sin x\, dx$$
$$du = dx \quad v = -\cos x$$

you find that

$$\int x \sin x\, dx = -x\cos x - \int -\cos x\, dx$$
$$= -x\cos x + \sin x + C$$

hence, $\int_{\pi/3}^{\pi/2} x \sin x\, dx = \left[-x\cos x + \sin x\right]_{\pi/3}^{\pi/2}$

$$= \left[\left(-\frac{\pi}{2}\right)\left(\cos\frac{\pi}{2}\right) + \sin\frac{\pi}{2}\right] - \left[\left(-\frac{\pi}{3}\right)\left(\cos\frac{\pi}{3}\right) + \sin\frac{\pi}{3}\right.$$

$$= (0 + 1) - \left(-\frac{\pi}{6} + \frac{\sqrt{3}}{2}\right)$$

$$= 1 + \frac{\pi}{6} - \frac{\sqrt{3}}{2}$$

$$= \frac{6 - 3\sqrt{3} + \pi}{6}$$

**Example 38:** Evaluate $\int_1^e x^2 \ln x\, dx$.

Using integration by parts with

$$u = \ln x \text{ and } dv = x^2\, dx$$
$$du = \frac{1}{x}\, dx \quad v = \frac{1}{3}x^3$$

you find that

$$\int x^2 \ln x\, dx = \frac{1}{3}x^3 \ln x - \int\left(\frac{1}{3}x^3\right)\left(\frac{1}{x}\right)dx$$
$$= \frac{1}{3}x^3 \ln x - \frac{1}{3}\int x^2\, dx$$
$$= \frac{1}{3}x^3 \ln x - \frac{1}{3}\cdot\frac{1}{3}x^3 + C$$
$$= \frac{1}{3}x^3 \ln x - \frac{1}{9}x^3 + C$$

hence,
$$\int_1^e x^2 \ln x\, dx = \left[\frac{1}{3}x^3 \ln x - \frac{1}{9}x^3\right]_1^e$$
$$= \left[\frac{1}{3}(e)^3 \ln e - \frac{1}{9}(e)^3\right] - \left[\frac{1}{3}(1)^3 \ln 1 - \frac{1}{9}(1)^3\right]$$
$$= \left[\frac{1}{3}e^3 - \frac{1}{9}e^3\right] - \left[0 - \frac{1}{9}\right]$$
$$= \frac{2}{9}e^3 + \frac{1}{9}$$
$$= \frac{1}{9}(2e^3 + 1)$$

**Example 39:** Evaluate $\int_{\pi/4}^{\pi/2} \cot^4 x\, dx$.

$$\int \cot^4 x\, dx = \int \cot^2 x \cot^2 x\, dx$$
$$= \int \cot^2 x\,(\csc^2 x - 1)\, dx$$

$$= \int (\cot^2 x \csc^2 x - \cot^2 x)\, dx$$

$$= \int \cot^2 x \csc^2 x\, dx - \int \cot^2 x\, dx$$

$$= \int \cot^2 x \csc^2 x\, dx - \int (\csc^2 x - 1)\, dx$$

$$= -\frac{1}{3}\cot^3 x + \cot x + x + C$$

hence,
$$\int_{\pi/4}^{\pi/2} \cot^4 x\, dx = \left[-\frac{1}{3}\cot^3 x + \cot x + x\right]_{\pi/4}^{\pi/2}$$

$$= \left[0 + 0 + \frac{\pi}{2}\right] - \left[-\frac{1}{3} + 1 + \frac{\pi}{4}\right]$$

$$= \frac{\pi}{4} - \frac{2}{3} \text{ or } \frac{3\pi - 8}{12}$$

**Example 40:** Evaluate $\int_{-\pi/2}^{3\pi/2} \cos^2 4x\, dx$.

$$\int \cos^2 4x\, dx = \int \left(\frac{1 + \cos 8x}{2}\right) dx$$

$$= \frac{1}{2}x + \frac{1}{16}\sin 8x + C$$

hence,
$$\int_{-\pi/2}^{3\pi/2} \cos^2 4x\, dx = \left[\frac{1}{2}x + \frac{1}{16}\sin 8x\right]_{-\pi/2}^{3\pi/2}$$

$$= \left[\frac{1}{2}\left(\frac{3\pi}{2}\right) + \frac{1}{16}\sin 12\pi\right] - \left[\frac{1}{2}\left(-\frac{\pi}{2}\right) + \frac{1}{16}\sin (-$$

$$= \left[\frac{3\pi}{4} + 0\right] - \left[-\frac{\pi}{4} + 0\right]$$

$$= \frac{4\pi}{4}$$

$$= \pi$$

**Example 41:** Evaluate $\int_{-3}^{3} \frac{dx}{x^2 + 9}$.

Because the integrand contains the form $a^2 + x^2$,

let $x = a \tan \theta = 3 \tan \theta$

$dx = 3 \sec^2 \theta \, d\theta$

and $x^2 + 9 = 9 \sec^2 \theta$ (Figure 18).

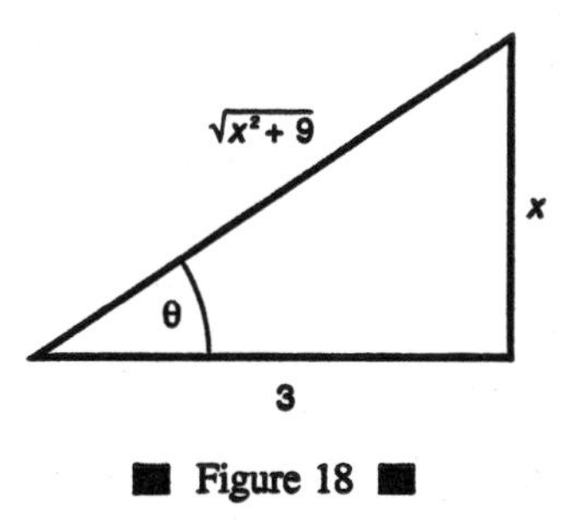


■ Figure 18 ■

Hence, $\int \frac{dx}{x^2 + 9} = \int \frac{3 \sec^2 \theta \, d\theta}{9 \sec^2 \theta}$

$$= \frac{1}{3} \int d\theta$$

$$= \frac{1}{3}\theta + C$$

$$= \frac{1}{3}\arctan\frac{1}{3}x + C$$

and $$\int_{-3}^{3}\frac{dx}{x^2+9} = \left[\frac{1}{3}\arctan\frac{1}{3}x\right]_{-3}^{3}$$

$$= \frac{1}{3}[\arctan 1 - \arctan(-1)]$$

$$= \frac{1}{3}\left[\frac{\pi}{4} - \left(-\frac{\pi}{4}\right)\right]$$

$$= \frac{1}{3}\left(\frac{\pi}{2}\right)$$

$$= \frac{\pi}{6}$$

**Example 42:** Evaluate $\int_{3}^{4}\frac{\sqrt{25-x^2}}{x}\,dx$.

Because the radical has the form $\sqrt{a^2 - x^2}$,

let $x = a\sin\theta = 5\sin\theta$

$dx = 5\cos\theta\, d\theta$

and $\sqrt{25 - x^2} = 5\cos\theta$ (Figure 19).

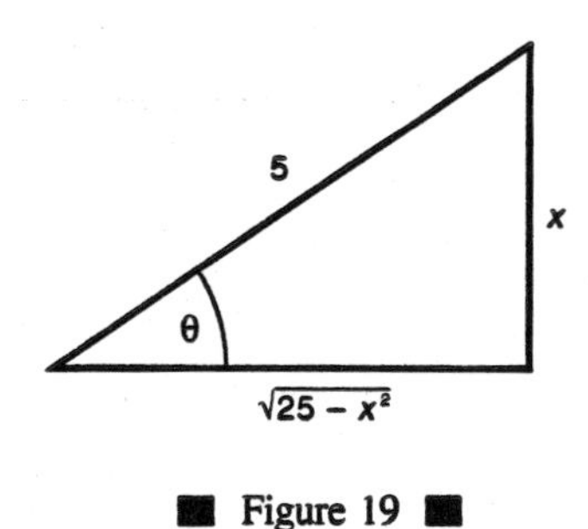


■ Figure 19 ■

Hence, $\int \frac{\sqrt{25 - x^2}}{x}\,dx = \int \frac{5\cos\theta}{5\sin\theta}(5\cos\theta\,d\theta)$

$$= 5\int \frac{\cos^2\theta}{\sin\theta}\,d\theta$$

$$= 5\int \frac{1 - \sin^2\theta}{\sin\theta}\,d\theta$$

$$= 5\int (\csc\theta - \sin\theta)\,d\theta$$

$$= -5\ln|\csc\theta + \cot\theta| + 5\cos\theta + C$$

$$= -5\ln\left|\frac{5}{x} + \frac{\sqrt{25 - x^2}}{x}\right| + 5\cdot\frac{\sqrt{25 - x^2}}{5} + C$$

$$= -5\ln\left|\frac{5 + \sqrt{25 - x^2}}{x}\right| + \sqrt{25 + x^2} + C$$

and $\int_3^4 \frac{\sqrt{25 - x^2}}{x}\, dx = \left[ -5 \ln \left| \frac{5 + \sqrt{25 - x^2}}{x} \right| + \sqrt{25 - x^2} \right]_3^4$

$$= [-5 \ln 2 + 3] - [-5 \ln 3 + 4]$$

$$= 5 (\ln 3 - \ln 2) - 1$$

$$= 5 \ln \frac{3}{2} - 1$$

The definite integral of a function has applications to many problems in calculus. Those considered in this section are areas bounded by curves, volumes by slicing, volumes of solids of revolution, and the lengths of arcs of a curve.

## Area

The area of a region bounded by the graph of a function, the $x$-axis, and two vertical boundaries can be determined directly by evaluating a definite integral. If $f(x) \geq 0$ on $[a,b]$, then the area ($A$) of the region lying below the graph of $f(x)$, above the $x$-axis, and between the lines $x = a$ and $x = b$ is

$$A = \int_a^b f(x)\,dx \qquad \text{(Figure 20)}$$

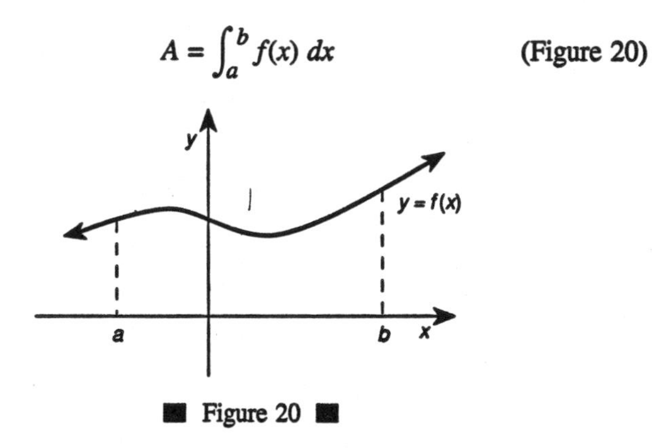


■ Figure 20 ■

If $f(x) \leq 0$ on $[a,b]$, then the area ($A$) of the region lying above the graph of $f(x)$, below the $x$-axis, and between the lines $x = a$ and $x = b$ is

$$A = \left| \int_a^b f(x)\,dx \right|$$

or $$A = -\int_a^b f(x)\,dx \qquad \text{(Figure 21)}$$

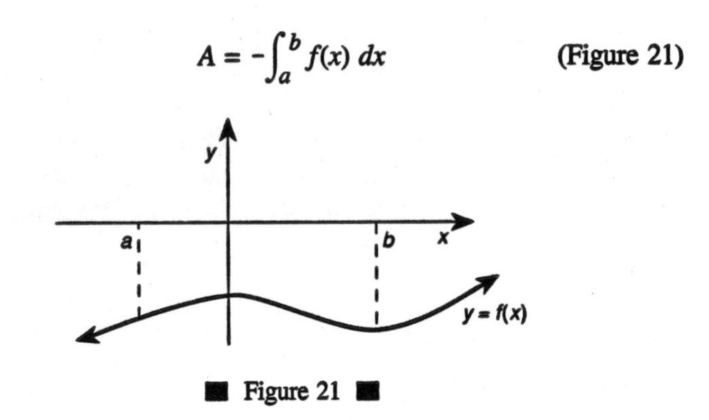


■ Figure 21 ■

If $f(x) \geq 0$ on $[a,c]$ and $f(x) \leq 0$ on $[c,b]$, then the area ($A$) of the region bounded by the graph of $f(x)$, the $x$-axis, and the lines $x = a$ and $x = b$ would be determined by the following definite integrals:

$$A = \left| \int_a^b f(x)\,dx \right|$$

$$A = \int_a^c f(x)\,dx - \int_c^b f(x)\,dx \qquad \text{(Figure 22)}$$

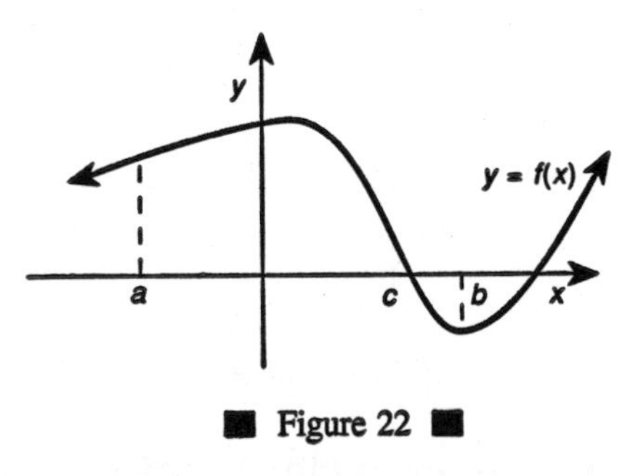


■ Figure 22 ■

Note that in this situation it would be necessary to determine all points where the graph of $f(x)$ crosses the $x$-axis and the sign of $f(x)$ on each corresponding interval.

For some problems that ask for the area of regions bounded by the graphs of two or more functions, it is necessary to determine the position of each graph relative to the graphs of the other functions of the region. The points of intersection of the graphs might need to be found in order to identify the limits of integration. As an example, if $f(x) \geq g(x)$ on $[a,b]$, then the area ($A$) of the region between the graphs of $f(x)$ and $g(x)$ and the lines $x = a$ and $x = b$ is

$$A = \int_a^b [f(x) - g(x)]\, dx \qquad \text{(Figure 23)}$$

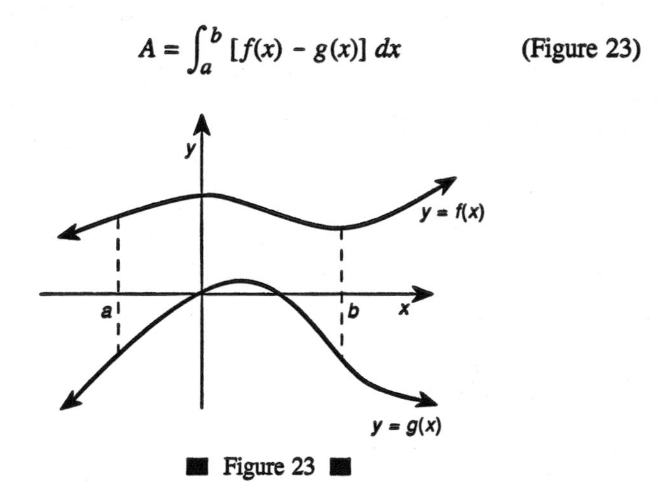


■ Figure 23 ■

Note that an analogous discussion could be given for areas determined by graphs of functions of $y$, the $y$-axis, and the lines $y = a$ and $y = b$.

**Example 1:** Find the area of the region bounded by $y = x^2$, the $x$-axis, $x = -2$, and $x = 3$.

Because $f(x) \geq 0$ on $[-2,3]$, the area ($A$) is

$$A = \int_{-2}^{3} x^2\, dx$$

$$= \frac{1}{3}x^3 \Big]_{-2}^{3}$$

$$= \frac{1}{3}(3)^3 - \frac{1}{3}(-2)^3$$

$$A = \frac{35}{3} \text{ or } 11\frac{2}{3}$$

**Example 2:** Find the area of the region bounded by $y = x^3 + x^2 - 6x$ and the $x$-axis.

Setting $y = 0$ to determine where the graph intersects the $x$-axis, you find that

$$x^3 + x^2 - 6x = 0$$

$$x(x^2 + x - 6) = 0$$

$$x(x + 3)(x - 2) = 0$$

$$x = 0,\ x = -3,\ x = 2$$

Because $f(x) \geq 0$ on $[-3,0]$ and $f(x) \leq 0$ on $[0,2]$ (Figure 24), the area ($A$) of the region is

$$A = \left|\int_{-3}^{2} (x^3 + x^2 - 6x)\,dx\right|$$

$$= \int_{-3}^{0} (x^3 + x^2 - 6x)\,dx - \int_{0}^{2} (x^3 + x^2 - 6x)\,dx$$

$$= \left[\frac{1}{4}x^4 + \frac{1}{3}x^3 - 3x^2\right]_{-3}^{0} - \left[\frac{1}{4}x^4 + \frac{1}{3}x^3 - 3x^2\right]_{0}^{2}$$

$$= \frac{63}{4} - \left(-\frac{16}{3}\right)$$

$$A = \frac{253}{12} \text{ or } 21\frac{1}{12}$$

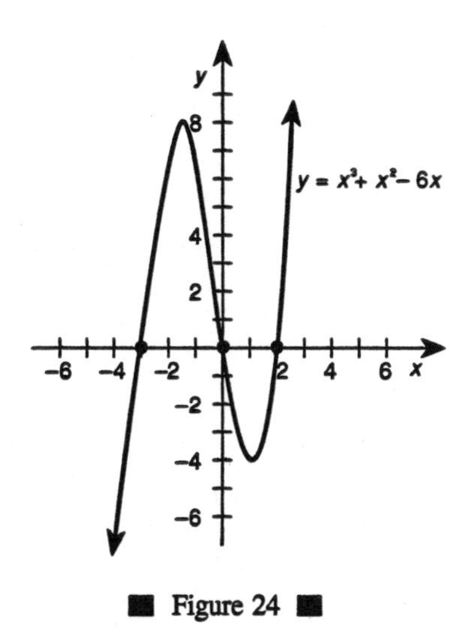


■ Figure 24 ■

**Example 3:** Find the area bounded by $y = x^2$ and $y = 8 - x^2$.

Because $y = x^2$ and $y = 8 - x^2$, you find that

$$\begin{aligned} x^2 &= 8 - x^2 \\ 2x^2 - 8 &= 0 \\ 2(x^2 - 4) &= 0 \\ 2(x + 2)(x - 2) &= 0 \\ x = -2, \quad x &= 2 \end{aligned}$$

hence, the curves intersect at $(-2,4)$ and $(2,4)$. Because $8 - x^2 \geq x^2$ on $[-2,2]$ (Figure 25), the area ($A$) of the region is

$$A = \int_{-2}^{2} [(8 - x^2) - (x^2)]\, dx$$

$$= \int_{-2}^{2} (8 - 2x^2)\, dx$$

$$= 8x - \frac{2}{3}x^3 \Big]_{-2}^{2}$$

$$= \frac{32}{3} - \left(-\frac{32}{3}\right)$$

$$A = \frac{64}{3} \text{ or } 21\frac{1}{3}$$

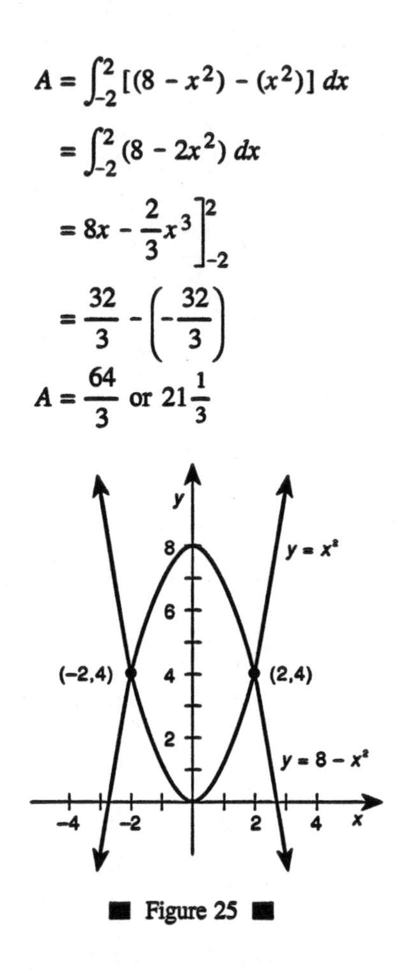


■ Figure 25 ■

## Volumes of Solids with Known Cross Sections

The definite integral can be used to find the volume of a solid with specific cross sections on an interval, provided we know a formula for the region determined by each cross section. If the cross sections generated are perpendicular to the $x$-axis, then their areas will be functions of $x$, denoted by $A(x)$. The volume ($V$) of the solid on the interval $[a,b]$ is

$$V = \int_a^b A(x)\, dx$$

If the cross sections are perpendicular to the $y$-axis, then their areas will be functions of $y$, denoted by $A(y)$. In this case, the volume {$V$) of the solid on $[a,b]$ is

$$V = \int_a^b A(y)\, dy$$

**Example 4:** Find the volume of the solid whose base is the region inside the circle $x^2 + y^2 = 9$ if cross sections taken perpendicular to the $y$-axis are squares.

Because the cross sections are squares perpendicular to the $y$-axis, the area of each cross section should be expressed as a function of $y$. The length of the side of the square is determined by two points on the circle $x^2 + y^2 = 9$ (Figure 26).

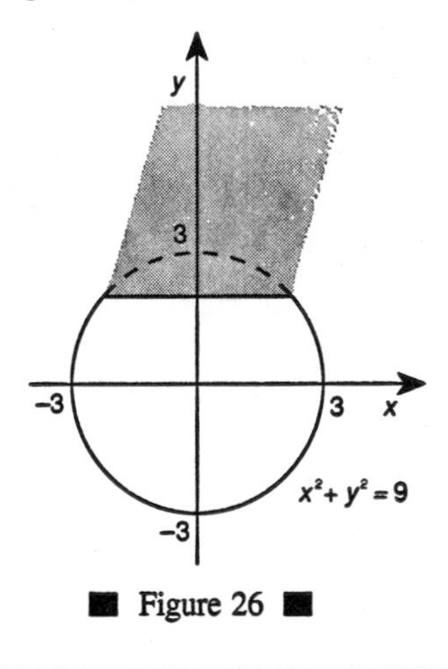


■ Figure 26 ■

The area ($A$) of an arbitrary square cross section is $A = s^2$, where $s = 2\sqrt{9 - y^2}$; hence,

$$A(y) = \left[2\sqrt{9 - y^2}\right]^2$$

$$A(y) = 4(9 - y^2)$$

The volume ($V$) of the solid is

$$V = \int_{-3}^{3} 4(9 - y^2)\, dy$$

$$= 4\left[9y - \frac{1}{3}y^3\right]_{-3}^{3}$$

$$= 4\left[18 - (-18)\right]$$

$$V = 144$$

**Example 5:** Find the volume of the solid whose base is the region bounded by the lines $x + 4y = 4$, $x = 0$, and $y = 0$, if cross sections taken perpendicular to the $x$-axis are semicircles.

Because the cross sections are semicircles perpendicular to the $x$-axis, the area of each cross section should be expressed as a function of $x$. The diameter of the semicircle is determined by a point on the line $x + 4y = 4$ and a point on the $x$-axis (Figure 27).

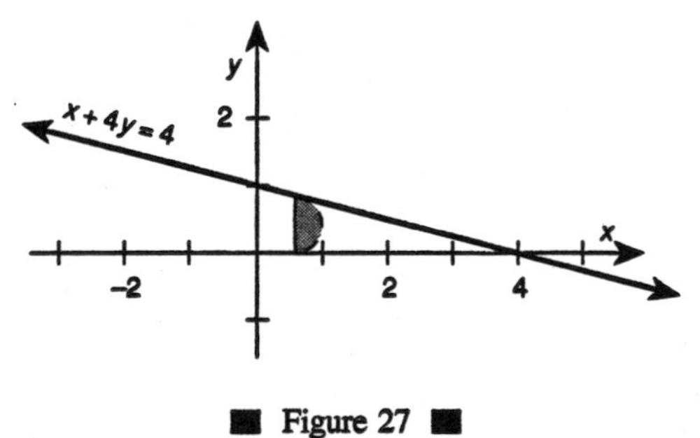

Figure 27

The area ($A$) of an arbitrary semicircle cross section is

$$A = \frac{1}{2}\pi r^2 = \frac{1}{2}\pi\left(\frac{1}{2}d\right)^2$$

where $d = \dfrac{4 - x}{4}$ and $r = \dfrac{4 - x}{8}$

hence,

$$A(x) = \frac{1}{2}\pi\left(\frac{4 - x}{8}\right)^2$$

$$A(x) = \frac{1}{128}\pi(4 - x)^2$$

The volume ($V$) of the solid is

$$V = \int_0^4 \frac{1}{128}\pi(4 - x)^2\,dx$$

$$= \frac{1}{128}\pi\int_0^4 (16 - 8x + x^2)\,dx$$

$$= \frac{1}{128}\pi\left[16x - 4x^2 + \frac{1}{3}x^3\right]_0^4$$

$$= \frac{1}{128}\pi\left[\frac{64}{3}\right]$$

$$V = \frac{\pi}{6}$$

## Volumes of Solids of Revolution

The definite integral can also be used to find the volume of a solid that is obtained by revolving a plane region about a horizontal or vertical line that does not pass through the plane. This type of solid will be made up of one of three types of elements—disks, washers, or cylindrical shells—each of which requires a different approach in setting up the definite integral to determine its volume.

**Disk method.** If the axis of revolution is the boundary of the plane region and the cross sections are taken perpendicular to the axis of revolution, then the **disk method** will be used to find the volume of the solid. Because the cross section of a disk is a circle with area $\pi r^2$, the volume of each disk is its area times its thickness. If a disk is perpendicular to the $x$-axis, then its radius should be expressed as a function of $x$. If a disk is perpendicular to the $y$-axis, then its radius should be expressed as a function of $y$.

The volume ($V$) of a solid generated by revolving the region bounded by $y = f(x)$ and the $x$-axis on the interval $[a,b]$ about the $x$-axis is

$$V = \int_a^b \pi\,[f(x)]^2\,dx$$

If the region bounded by $x = f(y)$ and the $y$-axis on $[a,b]$ is revolved about the $y$-axis, then its volume ($V$) is

$$V = \int_a^b \pi [f(y)]^2 \, dy$$

Note that $f(x)$ and $f(y)$ represent the radii of the disks or the distance between a point on the curve to the axis of revolution.

**Example 6:** Find the volume of the solid generated by revolving the region bounded by $y = x^2$ and the $x$-axis on $[-2,3]$ about the $x$-axis.

Because the $x$-axis is a boundary of the region, the disk method can be used (Figure 28).

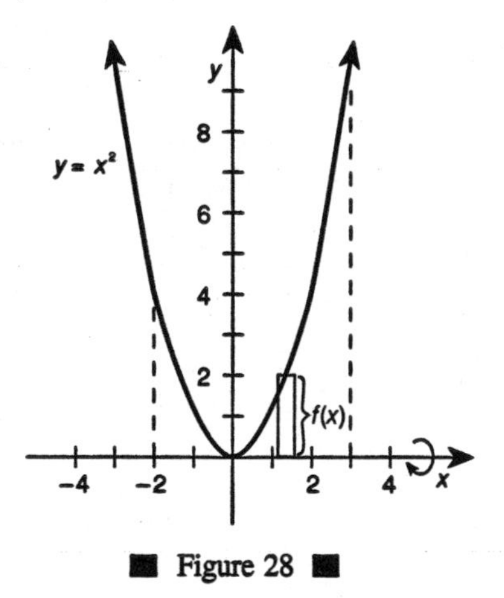


■ Figure 28 ■

The volume ($V$) of the solid is

$$V = \int_{-2}^{3} \pi (x^2)^2 \, dx$$

$$= \pi \int_{-2}^{3} x^4 \, dx$$

$$= \pi \left[ \frac{1}{5} x^5 \right]_{-2}^{3}$$

$$= \pi\left[\frac{243}{5} - \left(\frac{-32}{5}\right)\right]$$
$$V = 55\pi$$

**Washer method.** If the axis of revolution is not a boundary of the plane region and the cross sections are taken perpendicular to the axis of revolution, then the **washer method** will be used to find the volume of the solid. Think of the washer as a "disk with a hole in it" or as a "disk with a disk removed from its center." If $R$ is the radius of the outer disk and $r$ is the radius of the inner disk, then the area of the washer is $\pi R^2 - \pi r^2$, and its volume would be its area times its thickness. As noted in the discussion of the disk method, if a washer is perpendicular to the $x$-axis, then the inner and outer radii should be expressed as functions of $x$. If a washer is perpendicular to the $y$-axis, then the radii should be expressed as functions of $y$.

The volume ($V$) of a solid generated by revolving the region bounded by $y = f(x)$ and $y = g(x)$ on the interval $[a,b]$, where $f(x) \geq g(x)$, about the $x$-axis is

$$V = \int_a^b \pi\{[f(x)]^2 - [g(x)]^2\}\,dx$$

If the region bounded by $x = f(y)$ and $x = g(y)$ on $[a,b]$, where $f(y) \geq g(y)$ is revolved about the $y$-axis, then its volume ($V$) is

$$V = \int_a^b \pi\{[f(y)]^2 - [g(y)]^2\}\,dy$$

Note again that $f(x)$ and $g(x)$ and $f(y)$ and $g(y)$ represent the outer and inner radii of the washers or the distance between a point on each curve to the axis of revolution.

**Example 7:** Find the volume of the solid generated by revolving the region bounded by $y = x^2 + 2$ and $y = x + 4$ about the $x$-axis.

Because $y = x^2 + 2$ and $y = x + 4$, you find that

$$x^2 + 2 = x + 4$$
$$x^2 - x - 2 = 0$$
$$(x + 1)(x - 2) = 0$$
$$x = -1, \quad x = 2$$

The graphs will intersect at $(-1,3)$ and $(2,6)$ with $x + 4 \geq x^2 + 2$ on $[-1,2]$ (Figure 29).

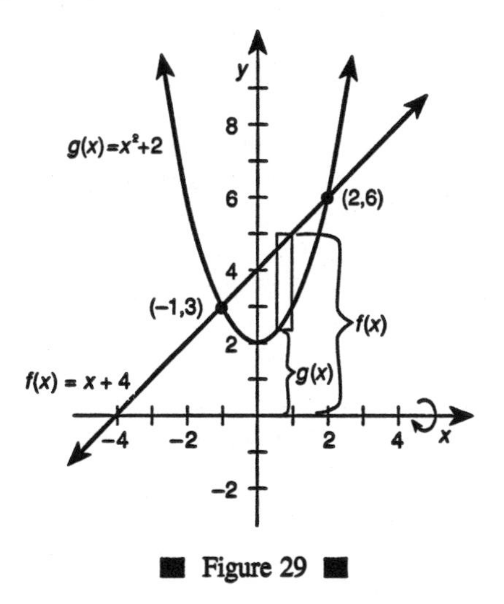


■ Figure 29 ■

Because the $x$-axis is not a boundary of the region, the washer method can be used and the volume ($V$) of the solid is

$$
\begin{aligned}
V &= \int_{-1}^{2} \pi\,[(x+4)^2 - (x^2+2)^2]\,dx \\
&= \int_{-1}^{2} \pi\,[(x^2+8x+16) - (x^4+4x^2+4)]\,dx \\
&= \pi \int_{-1}^{2} (-x^4 - 3x^2 + 8x + 12)\,dx \\
&= \pi \left[ -\frac{1}{5}x^5 - x^3 + 4x^2 + 12x \right]_{-1}^{2} \\
&= \pi \left[ \frac{128}{5} - \left( -\frac{34}{5} \right) \right] \\
V &= \frac{162\pi}{5}
\end{aligned}
$$

**Cylindrical shell method.** If the cross sections of the solid are taken parallel to the axis of revolution, then the **cylindrical shell method** will be used to find the volume of the solid. If the cylindrical shell has radius $r$ and height $h$, then its volume would be $2\pi rh$ times its thickness. Think of the first part of this product, $(2\pi rh)$, as the area of the rectangle formed by cutting the shell perpendicular to its radius and laying it out flat. If the axis of revolution is vertical, then the radius and height should be expressed in terms of $x$. If, however, the axis of revolution is horizontal, then the radius and height should be expressed in terms of $y$.

The volume ($V$) of a solid generated by revolving the region bounded by $y = f(x)$ and the $x$-axis on the interval $[a,b]$, where $f(x) \geq 0$, about the $y$-axis is

$$V = \int_a^b 2\pi x\, f(x)\, dx$$

If the region bounded by $x = f(y)$ and the $y$-axis on the interval $[a,b]$, where $f(y) \geq 0$, is revolved about the $x$-axis, then its volume ($V$) is

$$V = \int_a^b 2\pi y\, f(y)\, dy.$$

Note that the $x$ and $y$ in the integrands represent the radii of the cylindrical shells or the distance between the cylindrical shell and the axis of revolution. The $f(x)$ and $f(y)$ factors represent the heights of the cylindrical shells.

**Example 8:** Find the volume of the solid generated by revolving the region bounded by $y = x^2$ and the $x$-axis on $[1,3]$ about the $y$-axis.

In using the cylindrical shell method, the integral should be expressed in terms of $x$ because the axis of revolution is vertical. The radius of the shell is $x$, and the height of the shell is $f(x) = x^2$ (Figure 30).

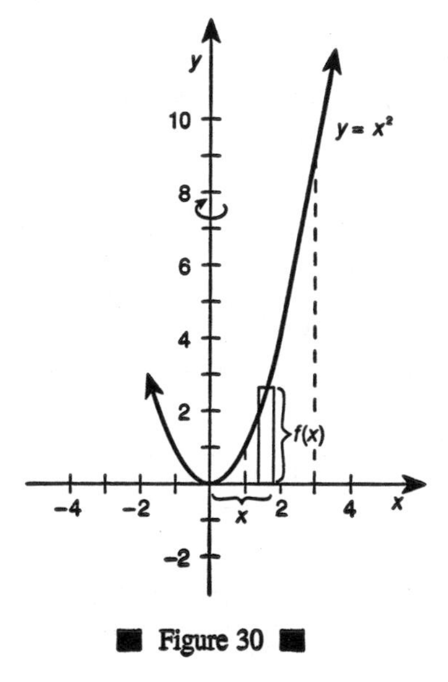


■ Figure 30 ■

The volume ($V$) of the solid is

$$V = \int_1^3 2\pi x \cdot x^2 \, dx$$
$$= 2\pi \int_1^3 x^3 \, dx$$
$$= 2\pi \left[\frac{1}{4}x^4\right]_1^3$$
$$= \frac{1}{2}\pi (81 - 1)$$
$$V = 40\pi$$

## Arc Length

The length of an arc along a portion of a curve is another application of the definite integral. The function and its derivative must both be continuous on the closed interval being considered for such an arc length to be guaranteed. If $y = f(x)$ and $y' = f'(x)$ are continuous on the closed interval $[a,b]$, then the arc length ($L$) of $f(x)$ on $[a,b]$ is

$$L = \int_a^b \sqrt{1 + [f'(x)]^2} \, dx$$

Similarly, if $x = f(y)$ and $x' = f'(y)$ are continuous on the closed interval $[a,b]$, then the arc length ($L$) of $f(y)$ on $[a,b]$ is

$$L = \int_a^b \sqrt{1 + [f'(y)]^2} \, dy$$

**Example 9: Find the arc length of the graph of $f(x) = \frac{1}{3}x^{3/2}$ on the interval [0,5].**

Because $$f(x) = \frac{1}{3}x^{3/2}$$

$$f'(x) = \frac{1}{2}x^{1/2}$$

and $$L = \int_0^5 \sqrt{1 + \left(\frac{1}{2}x^{1/2}\right)^2}\, dx$$

$$= \int_0^5 \sqrt{1 + \frac{1}{4}x}\, dx$$

$$= \int_0^5 \left(1 + \frac{1}{4}x\right)^{1/2} dx$$

$$= \left[4 \cdot \frac{2}{3}\left(1 + \frac{1}{4}x\right)^{3/2}\right]_0^5$$

$$= \frac{8}{3}\left(\frac{27}{8} - 1\right)$$

$$L = \frac{19}{3}$$

**Example 10:** Find the arc length of the graph of $f(x) = \ln(\sin x)$ on the interval $[\pi/4, \pi/2]$.

Because $$f(x) = \ln(\sin x)$$

$$f'(x) = \frac{\cos x}{\sin x} = \cot x$$

and

$$L = \int_{\pi/4}^{\pi/2} \sqrt{1 + \cot^2 x}\, dx$$

$$= \int_{\pi/4}^{\pi/2} \sqrt{\csc^2 x}\, dx$$

$$= \int_{\pi/4}^{\pi/2} \csc x\, dx$$

$$= \left[ -\ln |\csc x + \cot x| \right]_{\pi/4}^{\pi/2}$$

$$= (-\ln 1) - \left( -\ln \left|\sqrt{2} + 1\right| \right)$$

$$= 0 + \ln \left|\sqrt{2} + 1\right|$$

$$L = \ln \left|\sqrt{2} + 1\right|$$

$$L \approx 0.8813736$$